AF767639

Laser Plasma Accelerator for Ions

WSPC-PKUSP Physics Series
for Postgraduates – Vol. 2

Laser Plasma Accelerator for Ions

Xueqing Yan
Peking University, China

北京大学出版社
PEKING UNIVERSITY PRESS

World Scientific

Published by

World Scientific Publishing Co. Pte. Ltd.
5 Toh Tuck Link, Singapore 596224
USA office: 27 Warren Street, Suite 401-402, Hackensack, NJ 07601
UK office: 57 Shelton Street, Covent Garden, London WC2H 9HE

Library of Congress Control Number: 2025946672

British Library Cataloguing-in-Publication Data
A catalogue record for this book is available from the British Library.

激光离子加速物理及应用
Originally published in Chinese by Peking University Press © Peking University Press, 2023

LASER PLASMA ACCELERATOR FOR IONS

Copyright © 2026 by World Scientific Publishing Co. Pte. Ltd.

All rights reserved.

ISBN 978-981-98-2372-7 (hardcover)
ISBN 978-981-98-2373-4 (ebook for institutions)
ISBN 978-981-98-2374-1 (ebook for individuals)

For any available supplementary material, please visit
https://www.worldscientific.com/worldscibooks/10.1142/14602#t=suppl

Preface (I)

The electric field gradient of a laser accelerator is more than three orders of magnitude higher than that of a conventional radio-frequency accelerator, which can significantly reduce the size and cost of large accelerators. During the interaction between a laser and a solid target or a critical-density target, transient ultrashort ion beams and high-brightness gamma photons with a density ten orders of magnitude higher than that of existing accelerator beams can be produced. These have the potential to provide a new research tool for medical physics, nuclear physics, and particle physics. In-depth research on laser accelerators is expected to spark a new scientific and technological revolution, profoundly impacting disciplines such as particle and nuclear physics, astrophysics, as well as life and materials sciences.

The concept of laser acceleration was first proposed by American physicists Tajima and Dawson in 1979. In 2006, Leemans and colleagues demonstrated the feasibility of GeV monoenergetic electron acceleration, marking a milestone in this field. In China, research institutions such as the Shanghai Institute of Optics and Fine Mechanics of the Chinese Academy of Sciences (CAS), the China Academy of Engineering Physics, and the Institute of Physics of CAS, took the lead in conducting research on laser acceleration. Subsequently, numerous universities also initiated studies on the theory, experiments, and applications of laser acceleration. With further support and promotion from the High Energy Density Physics Professional Committee and the Particle Accelerator Branch of the Chinese Physical Society, an increasing number of young scholars and students have been attracted to this emerging field. Laser acceleration and its applications represent a branch of applied physics, involving many different disciplines, such as laser physics, plasma, accelerators, control, materials, and biomedicine. Currently, there is no systematic textbook on laser ion acceleration in China. Young scholars newly entering this field usually need to review a large number of documents to understand the basic knowledge and cutting-edge progress in this area. This book systematically introduces the basic theory of laser-plasma interactions and numerical simulation methods, acceleration physics mechanisms, ultrashort and ultraintense lasers, preparation of acceleration targets, beam transport and diagnostics of accelerators, and beam applications, which can help young scholars quickly get started.

Peking University has long been focused on new acceleration principles and methods. Since the beginning of this century, the Heavy Ion Research Institute has initiated explorations in laser particle acceleration, followed by theoretical research, experimental exploration, and application research. A laser proton accelerator facility has been built, which has already served the construction of Huairou Science City. The content of this book is both a summary of Peking University's laser accelerator R&D work over the past 20 years and of great significance to the development of accelerator physics. Next year, we will celebrate the 110th anniversary of the establishment of Peking University's physics discipline and the 40th anniversary of the founding of the Heavy Ion Research Institute. Here, I sincerely hope that the publication of this book will make an important contribution to the teaching and research work of accelerator science in our country.

Jiaer Chen
June 15, 2022

Preface (II)

Professor Yan Xueqing from the School of Physics at Peking University recently called me and informed me that he is currently leading the compilation of "Physics and Applications of Laser Ion Acceleration" as a university physics textbook, and he hoped that I would write a preface for it. Thus, I gladly took up my pen to record my thoughts and feelings.

The 2018 Nobel Prize in Physics was awarded to Professors Mourou and Strickland for their invention of chirped pulse amplification (CPA) technology. Since the invention of CPA technology, the focused intensity of ultrashort and ultraintense lasers has increased by eight orders of magnitude, and its applications in various scientific research and practical life have become much broader. One important application area is particle acceleration. Particle acceleration is one of the universal natural phenomena in the universe, and last century, scientists achieved fruitful results in the research of conventional accelerator physics and its applications. Statistics show that since the establishment of the Nobel Prize, nearly half of the awarded projects in physics have been related to particle acceleration research. Therefore, it can be said that the study of particle acceleration has driven the progress of human technological civilization.

The concept of laser wakefield acceleration was proposed by American physicists Tajima and Dawson in 1979. Since then, many research institutions both domestically and internationally have conducted research on laser acceleration. I myself served as the chief scientist leading the research of the "Ultraintense Laser-Driven Particle Acceleration and Its Applications" super 973 project from 2013 to 2018, which significantly accelerated the transformation process in China from "laser acceleration mechanism research" to "laser accelerator research." In recent years, with the support of the High Energy Physics Branch of the Chinese Physical Society, the Plasma Acceleration Professional Committee has been established. Under the auspices of the Chinese Physical Society, the High Energy Density Physics Professional Committee has also included laser acceleration as an important topic in the High Energy Density Physics Youth Forum.

Laser acceleration and its applications are a typical emerging interdisciplinary field involving lasers, plasmas, accelerators, control, materials, energy, and biomedicine. Much of the knowledge is scattered across multiple reference books and various recent publications. It is difficult for university students to obtain a systematic understanding of laser acceleration and its applications. Young scholars newly entering this field are also eager to read textbook-style books to gain a comprehensive understanding of the basic concepts, methods, and theories of laser acceleration and its applications. This is the background and significance of the compilation of this textbook.

"Physics and Applications of Laser Ion Acceleration" provides a concise and systematic introduction to the basic theory of laser-plasma interactions and numerical simulation methods, acceleration physics mechanisms, ultrashort and ultraintense lasers, preparation of acceleration targets, beam transport and diagnostics of accelerators, and transient beam applications, reflecting the latest progress in this field. More importantly, this book is a summary and showcase of the research achievements of Chinese scientists in this field over the years, documenting the process of China's growing research strength and influence in laser acceleration. Therefore, it holds significant value for the development of both accelerator physics and high energy density physics disciplines.

2023 marks the 110th anniversary of the beginning of physics education at Peking University, and the publication of this book will also serve as a precious birthday gift for the School of Physics at Peking University. As an old friend of Peking University, and also the former president of Shanghai Jiao Tong University and the chairman of the Chinese Physical Society, I would like to extend my advance congratulations on the 110th birthday of Peking University's physics discipline.

Thus, I write this preface to share with the readers.

Jie Zhang
June 15, 2022

Preface

In 2004, a milestone breakthrough was achieved in laser wakefield electron acceleration. Due to its significant importance in high-energy physics and applications, Mourou and Strickland were awarded the Nobel Prize in Physics in 2018.

With the development of laser technology, research on laser-driven particle acceleration has made remarkable progress. Many universities in China have researchers engaged in plasma acceleration studies, and the research results are increasingly attracting attention. Recently, Beijing Municipality approved the establishment of the "Beijing Laser Acceleration Innovation Center" in Huairou Science City, led by Peking University, to promote the application of laser acceleration in medicine, nuclear science, energy, and high energy density physics, laying the foundation for the future national major scientific infrastructure "Beijing Laser Accelerator Facility." The Particle Accelerator Branch of the Chinese Physical Society has also established a "Plasma Acceleration" professional committee. Together with the High Energy Density Physics Professional Committee, they have organized and initiated the "High Energy Density Physics Youth Scientist Forum" series of conferences. In recent years, the number of participants has reached 300–400 people (the first few editions were held at Peking University, Tsinghua University, Shanghai Jiao Tong University, University of Science and Technology of China, FAST in Guizhou, and National University of Defense Technology), and it continues to grow, representing a vibrant and cutting-edge interdisciplinary direction.

Universities and research institutes have also successively launched courses on laser acceleration or laser-plasma interactions. For example, the School of Physics at Peking University has already offered professional courses for undergraduate and graduate students, such as "Quantum Beam

Physics," "Fundamentals of Accelerator Physics," "Introduction to Laser Acceleration Principles," "Laser-Plasma Interactions," and "Introduction to High-Brightness X-Ray Sources." However, up to now, there has been no suitable textbook. The relevant knowledge is scattered across various reference books and the latest publications, and students often lack a systematic channel for acquiring knowledge.

In 2020, due to the sudden outbreak of the pandemic, students could only stay at home and we could not conduct normal offline teaching, which further highlighted the teaching issues. Therefore, I decided to organize the rapid writing of a series of professional textbooks for senior undergraduate and graduate students to change this situation and better introduce the latest knowledge of laser acceleration and related applications to them. This book is the first textbook in this series, introducing the basic knowledge and research methods of laser-plasma interactions, with a focus on ion acceleration research. Subsequently, it is hoped that other textbooks in related directions can be launched.

Looking back 20 years ago, the accelerator discipline at the university was facing difficulties such as a shortage of personnel and insufficient funding, and it was urgent to find a new research direction. Professor Guo Zhiyu, then the director of the Heavy Ion Research Institute, invited Professor Sheng Zhengming to serve as a visiting professor at Peking University to help the university conduct research on advanced acceleration. Since then, the Peking University team gradually began research on laser ion acceleration, starting from scratch to propose theories, build prototypes, and gradually initiate the application and promotion of laser accelerators. Under the concerted efforts of all team members, the CLAPA I facility was built, and the CLAPA II facility is also designed and about to be constructed at the Huairou Comprehensive National Research Center. These efforts have provided the source and firsthand materials for the writing of this book.

On the occasion of the imminent publication of this book, I am filled with emotions. First and foremost, I would like to express my sincere gratitude to Professor Chen Jia'er. Under his encouragement and inspiration of the "top-notch and down-to-earth" spirit, the new direction gradually developed. This would not have been possible without the efforts of the many colleagues who subsequently joined the CLAPA Laboratory, including Lin Chen, Ma Wenjun, Lu Haiyang, Zhu Kun, Zhao Yanying, Geng Yixing, and Yuan Zhongxi, among others. Among them, those who have directly or indirectly participated in the writing of this book include Lin Chen,

Ma Wenjun, Zhao Yanying, Geng Yixing, and Zhu Kun. Some graduate students and postdoctoral fellows have also contributed to the writing of some chapters, including Wu Xuezhi, Wang Kedong, Shou Yinren, Tang Yuhui, Lv Jianfeng, Xu Tianqi, Li Yuzhe, Liu Zhipeng, Li Dongyu, Kong Defeng, Xu Shirui, Chen Shiyu, Qi Guijun, Xia Yuhui, Mei Zhisong, Peng Ziyang, Zhang Hui, Yan Yang, and Ge Huiling. Without them, it would not have been possible to complete this textbook in such a short time. Under the organization of Ph.D. student Wu Xuezhi, Xia Yuhui, Zhang Jianyao, Lv Jianfeng, and Ma Qianyi were also responsible for the editing of the text, formulas, and formatting of this book. The compilation of this book has also received strong support and assistance from Professor Yu Jinqing of Hunan University and Professor Chen Min of Shanghai Jiao Tong University, to whom I extend my heartfelt thanks.

Xueqing Yan
Beijing
July 5, 2022

Contents

Chapter 1

Introduction

Particle accelerators are scientific devices that artificially accelerate charged particles and are indispensable tools for understanding the microscopic world. Many of the major discoveries in basic research in the 20th century can be attributed to the construction and use of accelerators. The discovery of the "God particle" — the Higgs boson — led to the 2013 Nobel Prize in Physics, once again bringing particle accelerators into the spotlight.

Traditional particle accelerators can be categorized by particle energy [1] into low-energy accelerators (energy below 10^8 eV), medium-energy accelerators (energy between 10^8 and 10^9 eV), high-energy accelerators (energy between 10^9 and 10^{11} eV), and ultra-high-energy accelerators (energy above 10^{11} eV). Based on their operating principles, they can be divided into electrostatic accelerators, linear accelerators, electron induction accelerators, cyclotrons, synchrotrons, and colliders. The largest and highest-energy particle accelerator in the world is the Large Hadron Collider (LHC) located in Europe, as shown in Figure 1.1. This facility uses two proton beams with an energy of 6.8 TeV to collide, achieving a particle energy of up to 13.6 TeV [2].

Basic research and frontier exploration have driven continuous progress in particle accelerator technology, and traditional particle accelerators have achieved remarkable success. However, as research into fundamental physics and the mysteries of the universe deepens, scientists still crave particles with higher energies. For traditional particle accelerators, further increases in particle energy are limited by materials, resources, and economics. For ordinary dielectric materials, an accelerating electric field exceeding 10^8 V/m in vacuum will cause breakdown and destruction of the material, thus limiting the accelerating gradient to below 100 MV/m. In terms of spatial

Fig. 1.1 European Large Hadron Collider (LHC).

resources, the size of the accelerator increases with the particle energy, occupying a large amount of land. For example, the linear electron accelerator at the SLAC National Accelerator Laboratory in the United States is 3 km in length, while the aforementioned LHC in Europe has a circumference of 27 km. The proposed Circular Electron Positron Collider (CEPC) in China and the Future Circular Collider (FCC) in Europe have circumferences on the order of 100 km [3]. Fermi once predicted that if traditional accelerators were used to accelerate particles to the PeV level, the acceleration length would need to encircle the Earth. The size of these accelerators has already reached the limit of what people can bear and has led to extremely high engineering costs.

The large footprint and exorbitant costs are major bottlenecks for the development of traditional accelerators towards higher energies. Finding new acceleration methods to break through the limitations of traditional accelerating gradients is urgent. Laser Plasma Acceleration (LPA), proposed by Tajima and Dawson in 1979, is a new acceleration mechanism [4] that has seen rapid development in theory and experiments over the past few decades and holds promise for the next generation of high-gradient accelerators.

With the continuous development of laser technology, research on laser acceleration of charged particles has garnered significant interest. When an ultraintense laser interacts with plasma, it can excite a very strong plasma electric field (e.g., charge separation field) within the plasma. The electric field strength can reach the level of the non-relativistic wave-breaking field:

$$E_0 \text{ (unit: V/cm)} \approx 0.96 n_0 \text{ (unit: cm}^{-3}\text{)}. \tag{1.1.1}$$

When the plasma density $n_0 = 10^{18}$ cm^{-3}, $E_0 \approx 100$ GV/m, which is three orders of magnitude higher than that of traditional accelerators, providing conditions for the miniaturization of accelerators. In this book, this new type of particle accelerator based on ultraintense laser-plasma interactions is referred to as a laser plasma accelerator, or simply a laser accelerator.

Research in the field of laser particle acceleration is inseparable from the rapid development of laser technology. Since the birth of the world's first ruby solid-state laser in 1960 [5], the utilization of lasers and related scientific and technological developments have continuously ushered in new eras. Figure 1.2 [6] shows the changes in laser focal intensity over the years and the corresponding research objects in physics. The development of lasers has gone through several major technological innovations. Initially, free-running lasers had pulse widths of about several hundred μs and peak intensities of the kW level. After the emergence of Q-switching technology, laser pulse widths could be reduced to the ns (nanosecond, i.e., 10^{-9} s) timescale, with power reaching the MW (10^6 W) level. Subsequently, the use of mode-locking technology and broadband dyes enabled laser pulse widths to reach the ps (picosecond, i.e., 10^{-12} s) level, with peak power densities as high as the GW (10^9 W) [7, 8] level. At this point, the laser intensity exceeded 10^{13} W/cm^2, and atoms would undergo multiphoton ionization and tunnel ionization in strong lasers, bringing laser-matter interactions into the realm of atomic physics. In the 1970s, due to the potential of lasers in driving inertial confinement fusion (ICF), laser-plasma interactions began to be extensively studied. In the 1980s, proposals were made for accelerating protons in laser-driven strong magnetic fields [9], but

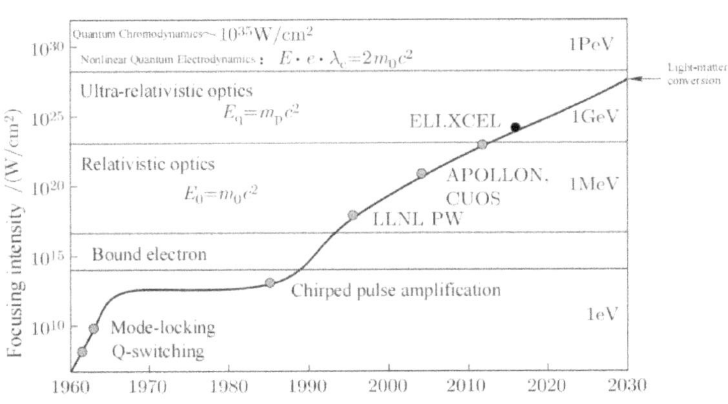

Fig. 1.2 The development of laser technology and the expansion of corresponding research fields [5].

during this stage, even with large lasers, ion energies remained below 1 MeV, awaiting the advent of ultraintense lasers.

With the discovery of the new titanium sapphire laser medium in 1982, the emergence of the Kerr-lens mode-locking technique in 1991, and especially the birth of chirped pulse amplification (CPA) technology in 1985 [10], ultrashort (several ps to tens of fs, where fs $= 10^{-15}$ s) and ultraintense lasers became a reality in the 1990s. CPA technology can concentrate laser energy into the fs timescale, with peak powers reaching the TW-PW (10^{12}–10^{15} W) level and intensities up to 10^{18}–10^{22} W/cm^2. When the laser intensity exceeds 10^{18} W/cm^2, the strength of its electromagnetic field can accelerate electrons to near-light speed ($v_{\text{osc}} \sim c$) within a single cycle, and the motion of electrons must be described considering relativistic effects. Laser-matter interactions thus entered the new field of relativistic laser plasma physics [11].

In 2000, Clark *et al.* [12] in the UK and Snavely *et al.* [13] in the US successively discovered that relativistically intense lasers striking solid targets could produce proton beams in the MeV range. Research on laser-driven ion acceleration has now exceeded 20 years. Table 1.1 summarizes the main laser-driven ion acceleration experimental parameters and results from 2000 to 2022, based on the review article by Daido *et al.* [14].

Table 1.1 Summary of laser-driven ion acceleration experiments.

Reference	Laser Energy (J)	Laser Pulse Width (fs)	Laser Intensity (W/cm^2)	Target Material and Thickness (μm)	Proton (Ion) Energy (MeV/ Nucleon)	Acceleration Mechanism
Clark *et al.* (2000)	50	1000	5×10^{19}	Al 125	18	TNSA
Snavely *et al.* (2000)	423	500	3×10^{20}	CH 100	58	TNSA
Krushelnick *et al.* (2000)	50	1000	5×10^{19}	Al 125	30	TNSA
Nemoto *et al.* (2001)	4	400	6×10^{18}	Mylar 6	10	TNSA
Mackinnon *et al.* (2002)	10	100	1×10^{20}	Al 3	24	TNSA
Patel *et al.* (2003)	10	100	5×10^{18}	Al 20	12	TNSA
Spencer *et al.* (2003)	0.2	60	7×10^{18}	Mylar 23	1.5	TNSA
Spencer *et al.* (2003)	0.2	60	7×10^{18}	Al 12	0.9	TNSA
McKenna *et al.* (2004)	233	700	2×10^{20}	Fe 100	40	TNSA

Table 1.1 (*cont'd*).

Reference	Laser Energy (J)	Laser Pulse Width (fs)	Laser Intensity (W/cm^2)	Target Material and Thickness (μm)	Proton (Ion) Energy (MeV/ Nucleon)	Acceleration Mechanism
Kaluza *et al.* (2004)	0.85	150	1.3×10^{19}	Al 20	4	TNSA
Oishi *et al.* (2005)	0.12	55	6×10^{18}	Cu 5	1.3	TNSA
Fuchs *et al.* (2006)	10	320	6×10^{19}	Al 20	20	TNSA
Neely *et al.* (2006)	0.3	33	1×10^{19}	Al 0.1	4	TNSA
Willingale *et al.* (2006)	340	1000	6×10^{20}	He jet 2000	10	TNSA
Ceccotti *et al.* (2007)	0.65	65	5×10^{18}	Mylar 0.1	5.25	TNSA
Robson *et al.* (2007)	310	1000	6×10^{20}	Al 10	55	TNSA
Robson *et al.* (2007)	160	1000	3.2×10^{20}	Al 10	38	TNSA
Robson *et al.* (2007)	30	1000	6×10^{19}	Al 10	16	TNSA
Antici *et al.* (2007)	1	320	1×10^{18}	Si_3N_4 0.03	7.3	TNSA
Yogo *et al.* (2007)	0.71	55	8×10^{18}	Cu 5	1.4	TNSA
Yogo *et al.* (2008)	0.8	45	1.5×10^{19}	Polyimide 7.5	3.8	TNSA
Nishiuchi *et al.* (2008)	1.7	34	3×10^{19}	Polyimide 7.5	4	TNSA
Flippo *et al.* (2008)	20	600	1.1×10^{19}	Flat-top conical Al 10	30	TNSA
Safronov *et al.* (2008)	6.5	900	1×10^{19}	Al 2	8	TNSA
Henig *et al.* (2009)	0.7	45	5×10^{19}	DLC 5.4 nm	13	RPA
Fukuda *et al.* (2009)	0.15	40	7×10^{17}	CO_2+He jet 2000	10	TNSA
Zeil *et al.* (2010)	3	30	1×10^{21}	Ti 2	17	TNSA
Gaillard *et al.* (2011)	82	670	1.5×10^{20}	Flat-top conical Cu 12.5	67.5	TNSA
Haberberger *et al.* (2012)	60	3000	6.5×10^{16}	Hydrogen gas jet	20	CSA
Margarone *et al.* (2012)	2	30	5×10^{19}	Nanospheres 0.535	8.6	TNSA
Kar *et al.* (2012)	200	700	3×10^{20}	Cu 0.1	7	RPA
Hegelich *et al.* (2013)	90	540	2×10^{20}	DLC 58 nm C	44	BOA
Zhang *et al.* (2015)	8.4	65	3.5×10^{19}	DLC 30 nm	4.7	CSA
Bin *et al.* (2015)	5	50	2×10^{20}	CNF 5+ DLC 20 nm	15	RPA
Wagner *et al.* (2015)	190	500	8×10^{20}	Polymer 0.75	61	BOA

Table 1.1 (*cont'd*).

Reference	Laser Energy (J)	Laser Pulse Width (fs)	Laser Intensity (W/cm^2)	Target Material and Thickness (μm)	Proton (Ion) Energy (MeV/ Nucleon)	Acceleration Mechanism
Margarone *et al.* (2015)	30	10	7×10^{20}	Nanospheres 0.72	30	TNSA
Wagner *et al.* (2016)	200	500	2.6×10^{20}	Plastic 0.9	85	TNSA
Passonl *et al.* (2016)	7.4	30	4.5×10^{20}	C foam 8+ Al 0.75	30	TNSA
Kim *et al.* (2016)	8.5	30	6.1×10^{20}	Polymer 15 nm	93	RPA
Zhang *et al.* (2017)	13	55	6.9×10^{19}	Plastic 40 nm	9	CSA
Scullion *et al.* (2017)	6	45	6×10^{20}	Carbon 10 nm C	25	RPA
Higginson *et al.* (2018)	210	900	3×10^{20}	Plastic 80 nm	$\sim$ 100	RPA-TNSA
Bin *et al.* (2019)	5	50	2×10^{20}	CNF 8+ DLC 20 nm	29	TNSA
Ma *et al.* (2019)	9.2	33	5.5×10^{20}	CNF 80+ DLC 20 nm C	48	RPA-TNSA
Vallieres *et al.* (2019)	220	700	5×10^{20}	Nano-particles+ Al 15	50	TNSA
Hong *et al.* (2020)	30	30	5×10^{20}	Carbon 1	21	TNSA
Dover *et al.* (2020)	10	40	5×10^{21}	Steel 5	30	TNSA
Qin *et al.* (2021)	28	35	1.8×10^{20}	Al 10	13.9	TNSA
Raffestin *et al.* (2021)	450	610	7.9×10^{18}	CH 50	51	TNSA
Ziegler *et al.* (2021)	18	30	5.4×10^{21}	Formvar 0.4	70	TNSA
Liu *et al.* (2021)	130	1000	1.5×10^{18}	Au 10	18.9	TNSA
Kuramitsu *et al.* (2022)	10	40	5×10^{21}	Graphene 8 nm	13.2	TNSA
Keppler *et al.* (2022)	17	120	3×10^{20}	Al 6.5	25	TNSA

After more than twenty years of research on laser ion acceleration, various ion acceleration mechanisms such as Target Normal Sheath Acceleration (TNSA), Radiation Pressure Acceleration (RPA), and Shock Acceleration (CSA) have been proposed successively. Experimentally, the proton energy generated by laser acceleration has approached 100 MeV. Compared with traditional accelerators, in addition to the advantage of small size due to the high accelerating gradient, the high-energy ion beams produced by lasers also have the characteristics of narrow pulse width, high instantaneous current, and good directionality. Such short-pulse, high-current ion beams have great potential application in many fields such as tumor therapy, electromagnetic field diagnostics, warm dense matter generation, and proton radiography, and are continuously being verified by experimental work. These applications will also provide new research tools for the development of disciplines such as magnetic confinement fusion, inertial confinement fusion, medical physics, nuclear physics, and high energy density physics. On the other hand, laser plasma accelerators are still in the experimental research stage, and there are still some physical problems that need to be solved urgently. For example, although the issues of ion energy spread, stability, and reliability have been preliminarily addressed in experiments, further efforts are needed to increase proton energy to above 200 MeV. To promote the application of laser particle accelerators as soon as possible, it is necessary to further increase the peak power and focusing quality of lasers, as well as to gain a deeper physical understanding and improvement of the laser plasma acceleration process.

This book is mainly aimed at young scholars who have just entered the field of laser plasma acceleration or are interested in high energy density physics. The book introduces the knowledge of laser ion acceleration from the aspects of theory, numerical simulation, experiments, and applications, with the hope that readers can systematically and comprehensively understand the field of laser ion acceleration through this book. Chapter 2 first introduces the relevant knowledge of numerical simulation, including the most commonly used Particle-in-Cell (PIC) and fluid simulations. Numerical simulation is an important mean of studying laser-plasma interactions at present. This chapter not only introduces the basic principles of numerical simulation research on laser-plasma interactions but also provides easy-to-use tutorials for the commonly used open-source PIC simulation software EPOCH and Smilei, as well as the fluid simulation software MULTI, to facilitate beginners. Chapter 3 introduces the theoretical foundations of laser-plasma interactions. Starting from the single-particle image, it introduces the interaction process of relativistic lasers with single particles, and then introduces the collective effects of plasma and the basic description methods of laser-plasma interactions. Chapter 4 introduces several major laser-driven ion acceleration mechanisms and combines experimental results to give readers a specific understanding of the current research progress in laser ion acceleration. Chapter 5 introduces several key factors in the Radiation Pressure Acceleration mechanism that limit acceleration energy and conversion efficiency, including the realization of ultra-high laser contrast, finite focal spot effects, and the development of transverse instabilities. Chapters 6 to 9 focus on various aspects of laser ion acceleration experiments, introducing knowledge related to laser technology, target preparation, ion acceleration experiments, diagnostics, and beam transport. The final chapter introduces some potential applications of laser-driven ion beams and secondary radiation generated during the acceleration process.

References

[1] Chen Jia'er. Fundamentals of Accelerator Physics [M]. Beijing: Peking University Press, 2012.
[2] ATLAS Collaboration. ATLAS event display of top-pair production in 13.6 TeV collisions during Run 3 [R]. 2022.
[3] `http://www.ihep.cas.cn/dkxzz/cepc/` [Z].
[4] Tajima T., Dawson J. M. Laser electron accelerator [J]. *Physical Review Letters*, 1979, 43(4): 267–270.
[5] Maiman T. H. Stimulated optical radiation in ruby [J]. *Nature*, 1960, 187(4736): 493–494.
[6] Mourou G. Nobel lecture: extreme light physics and application [J]. *Reviews of Modern Physics*, 2019, 91(3): 030501.
[7] Hargrove L. E., Fork R. L., Pollack M. A. Locking of He-Ne laser modes induced by synchronous intracavity modulation [J]. *Applied Physics Letters*, 1964, 5(1): 4–5.
[8] DiDomenico M. Small-signal analysis of internal (coupling-type) modulation of lasers [J]. *Journal of Applied Physics*, 1964, 35(10): 2870–2876.
[9] Forslund D. W., Brackbill J. U. Magnetic-field-induced surface transport on laser irradiated foils [J]. *Physical Review Letters*, 1982, 48(23): 1614–1617.
[10] Strickland D., Mourou G. Compression of amplified chirped optical pulses [J]. *Optics Communications*, 1985, 55(6): 447–449.
[11] Mourou G. A., Barty C. P. J., Perry M. D. Ultrahigh-intensity lasers: physics of the extreme on a tabletop [J]. *Physics Today*, 1998, 51(1): 22–28.
[12] Clark E. L., Krushelnick K., Davies J. R. *et al.* Measurements of energetic proton transport through magnetized plasma from intense laser interactions with solids [J]. *Physical Review Letters*, 2000, 84(4): 670–673.
[13] Snavely R. A., Key M. H., Hatchett S. P. *et al.* Intense high-energy proton beams from petawatt-laser irradiation of solids [J]. *Physical Review Letters*, 2000, 85(14): 2945–2948.
[14] Daido H., Nishiuchi M., Pirozhkov A. S. Review of laser-driven ion sources and their applications [J]. *Reports on Progress in Physics*, 2012, 75(5): 056401.

Chapter 2

Numerical Simulation of Laser-Plasma Interactions

When studying the interaction processes of laser-plasma, the main methods employed are theoretical analysis, experimental research, and numerical simulation. Among these, theoretical analysis starts from fundamental equations and, under reasonable assumptions and approximations, provides physical explanations for phenomena, generally applicable to linear processes. However, the interaction of ultraintense lasers with plasma involves many strongly nonlinear phenomena, such as collective interactions and instability development, which pose great difficulties for theoretical analysis methods. Experimental research is the foundation of physics, and experiments on laser-plasma interactions are vigorously carried out worldwide, yielding fruitful research results. Nevertheless, ultraintense and ultrashort laser systems are highly sophisticated and complex, and each round of experiments requires very thorough preparation. Additionally, there are numerous factors affecting experimental results, and controlling and tuning these uncertainties solely from an experimental perspective is extremely challenging, time-consuming, and labor-intensive. This is where the importance of numerical simulation comes into play. Due to the flexibility of numerical simulation, conditions that are difficult to achieve in experiments can be easily realized in numerical calculations, often helping us to identify the optimal experimental schemes. In fact, experiments and simulations often complement each other. On one hand, new discoveries in experiments can be further analyzed and explained through simulation work. Numerical simulation can comprehensively reproduce or dissect the physical processes in experiments, which helps to better establish physical models and understand experimental phenomena. On the other hand, many forward-looking simulation studies can put forward requirements and guidance for experiments, which are then verified by experiments. Many important discoveries are based on simulation work.

Numerical simulations of plasma can generally be divided into kinetic simulations and fluid simulations. Kinetic simulations consider a more detailed plasma model of particles under the influence of electromagnetic fields and are more suitable for describing the kinetic behavior of plasma. Fluid simulations, on the other hand, are often used to focus on physical processes on larger spatiotemporal scales, such as the pre-expansion process of solid targets or nanostructured targets under the action of laser prepulses. These will be introduced separately below.

2.1 Kinetic Simulation

Kinetic simulation includes two methods, namely solving the Vlasov or Fokker–Planck equations and the Particle-In-Cell (PIC) method. The PIC method, which calculates from the single-particle motion equations of plasma, is commonly referred to as the PIC method and is one of the main methods for numerical simulation and theoretical research work in laser acceleration studies. A brief introduction to this method will be given below.

2.1.1 *Particle-In-Cell (PIC) Method*

The Particle-In-Cell (PIC) method tracks the motion of a large number of charged particles in external and self-generated electromagnetic fields to obtain the macroscopic properties and motion laws of plasma. It is a very effective numerical experiment for studying plasma kinetic behavior [1–3]. It can provide real-time evolution of various physical phenomena, has the advantages of good reproducibility, intuitive and rich physical images, and is suitable for large-scale parameter studies, especially for highly nonlinear physical processes. The PIC method places high demands on computer performance. Conducting PIC simulations usually requires high-performance single machines or computer clusters, and physical operations with large scale or density spans often require large supercomputers. The rapid development of high-performance computing technology in recent decades has provided a good platform for numerical simulations in laser-plasma research.

In the PIC method, it is necessary to numerically solve the Maxwell equations for electromagnetic fields and the motion equations for particles, as follows:

$$\left.\begin{aligned}\nabla\cdot\boldsymbol{E}&=\frac{\rho}{\varepsilon_0},\\ \nabla\cdot\boldsymbol{B}&=0,\\ \nabla\times\boldsymbol{E}&=-\frac{\partial\boldsymbol{B}}{\partial t},\\ \nabla\times\boldsymbol{B}&=\mu_0\boldsymbol{J}+\mu_0\varepsilon_0\frac{\partial\boldsymbol{E}}{\partial t},\end{aligned}\right\}\tag{2.1.1}$$

$$\left.\begin{aligned}\frac{\mathrm{d}\boldsymbol{r}_j}{\mathrm{d}t}&=\boldsymbol{v}_j,\\ \frac{\mathrm{d}\boldsymbol{v}_j}{\mathrm{d}t}&=\frac{q_j}{m_j}(\boldsymbol{E}+\boldsymbol{v}_j\times\boldsymbol{B}),\end{aligned}\right\}\tag{2.1.2}$$

where

$$\rho=\sum_j q_j\boldsymbol{\delta}(\boldsymbol{r}_j-\boldsymbol{r}),\tag{2.1.3}$$

$$\boldsymbol{J}=\sum_j q_j\boldsymbol{v}_j\boldsymbol{\delta}(\boldsymbol{r}_j-\boldsymbol{r}).\tag{2.1.4}$$

Here, q_j, m_j, $\boldsymbol{r}_j$, $\boldsymbol{v}_j$ are the charge, mass, position, and velocity of the particle labeled j, respectively. c is the speed of light, $\boldsymbol{\delta}$ is the Dirac-Delta function, and $\boldsymbol{E}$, $\boldsymbol{B}$, ρ, and $\boldsymbol{J}$ represent the electric field, magnetic field, charge density, and current density, respectively.

The basic process of PIC simulation is shown in Figure 2.1 ($\boldsymbol{r}$ is represented as $\boldsymbol{x}$ in the figure): $(\boldsymbol{r}_j, \boldsymbol{v}_j) \to (\rho, \boldsymbol{J}) \to (\boldsymbol{E}, \boldsymbol{B}) \to (\boldsymbol{r}_j, \boldsymbol{v}_j) \to \cdots$.

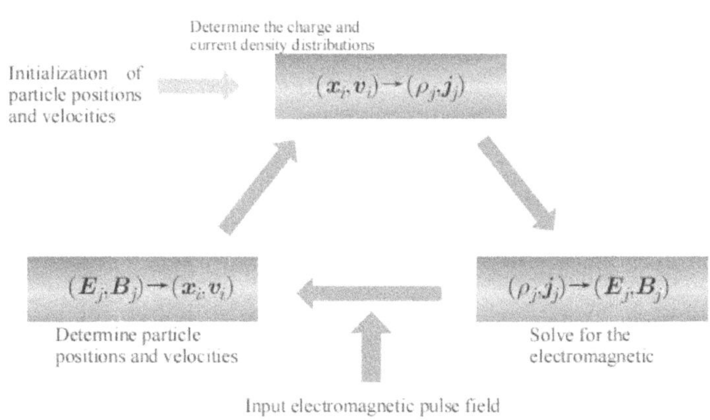

Fig. 2.1 The basic process of PIC simulation [4].

First, define the spatial grid and particles used in the simulation, and then perform the following loop:

(1) Distribute the charges of particles within the grid to the grid points to obtain the charge and current distributions.

The motion of charged particles generates current, which in turn produces the corresponding electromagnetic fields. According to the continuity equation, the change in particle charge density distribution can characterize the current density. In PIC simulations, the main attributes of particles, such as charge, are represented with the grid as the spatial unit and do not involve particle velocity, which is determined only by the given interpolation function and the position of the particles. Therefore, the local current density can be calculated by first solving the charge density components of particles at the nearby grid points at different times, and then obtaining the current density at the grid point from the rate of change of the grid point charge density.

(2) Calculate the electromagnetic fields by solving the electromagnetic field equations on the grid.

The electromagnetic fields are solved by solving Maxwell's equations. The Finite Difference Time Domain (FDTD [5, 6]) algorithm, which is a mature and commonly used method for solving Maxwell's equations in PIC, allocates the electric and magnetic fields to the positions of the grid points. The characteristic of the grid method is that the electromagnetic field at each grid point is only related to its adjacent grid points. After the Maxwell equations in the form of time-space partial differential equations are discretized, the evolution of the electromagnetic fields can be iterated by alternately updating the electric and magnetic fields at the time-space grid points.

(3) Solve the particle motion using the classical or relativistic particle motion equations.

The solution of particle motion and the solution of the Maxwell fields are carried out alternately. The motion of particles changes the field distribution, and the update of field information further promotes the motion of particles. In PIC, the relativistic particle motion equation is solved using the half-acceleration — rotation — half-acceleration method [7]. Simply put, the electric field force acts for half a time step first, then the magnetic field acts for one time step, and finally the electric field acts for the remaining half time step. This is also known as the "Boris method."

By iteratively solving in this manner, a description of the plasma motion patterns can be provided.

The actual number of particles in the plasma region far exceeds the capabilities of current computer simulations, making it impossible to simulate every single particle. Meanwhile, in the phase space of the plasma distribution function, the contributions of charged particles to the electromagnetic fields and the forces exerted on the particles by the fields are essentially the same around a point $(\boldsymbol{r}, \boldsymbol{v})$, and the motion patterns of these charged particles are also essentially identical. Therefore, a "macro-particle" can be used to represent these particles. For this reason, Buneman [8] and Dawson [9] proposed the "macro-particle" model. In the macro-particle system, the mass, charge, and density of a particle (representing N electrons) are $m' = Nm$, $e' = Ne$, and $n' = n/N$, respectively. This simplification usually does not affect the computational accuracy when describing collective effects in plasma but significantly reduces the computational load.

With the continuous application of PIC programs in the field of laser-plasma physics, many improvements and optimizations have been made to PIC codes, such as moving window techniques, higher-order interpolating particle cloud methods, multi-time-scale algorithms, variable-weight particles, and implicit simulation methods. At the same time, to meet different scientific research needs, traditional PIC codes have also added modules that implement specific computational functions, such as collision modules, ionization modules, radiation modules, and quantum dynamics modules that include the production of electron-positron pairs. After decades of development, the PIC method has become very mature, and many simulation results have been verified by experiments. Currently, several one- to three-dimensional fully electromagnetic relativistic particle simulation programs have been successively developed both domestically and internationally, including VLPL [10] developed by Pukhov at the Max Planck Institute for Quantum Optics in Germany, OSIRIS [11] developed by the University of California, Los Angeles (UCLA), VORPAL [12] developed by the University of Colorado, and KLAP [13] developed by Shanghai Jiao Tong University. Currently, open-source particle simulation programs mainly include the one-dimensional three-dimensional program LPIC++ [14] developed by the Max Planck Institute for Quantum Optics, the EPOCH [15] program developed by the CFSA group at the University of Warwick, and the Smilei [7] program developed by the LULI laboratory in France. These particle simulation programs are widely used in many fields involved in plasma physics, such as laser inertial confinement fusion, particle acceleration, space plasma, and free-electron lasers.

2.1.2 *Courant–Friedrichs–Lewy Condition*

In PIC simulations, the FDTD method is used to solve the electromagnetic fields. The electromagnetic fields are discretely distributed in the grid. Figure 2.2 shows the most commonly used Yee grid electromagnetic field distribution: the electric and magnetic fields are staggered by 1/2 grid length in each dimension. In the calculation, the two curl equations in Maxwell's equations are used to advance the evolution of the electromagnetic fields: the temporal evolution of the electric field depends on the spatial curl of the magnetic field, and the curl calculation in the simulation is obtained by differencing the field quantities at nearby spatial points. Therefore, to update the electric field, the distribution of the electric field and the nearby magnetic field at this moment is required, and the magnetic field calculation is similar. This grid definition is designed to enhance the accuracy of spatial gradient calculations.

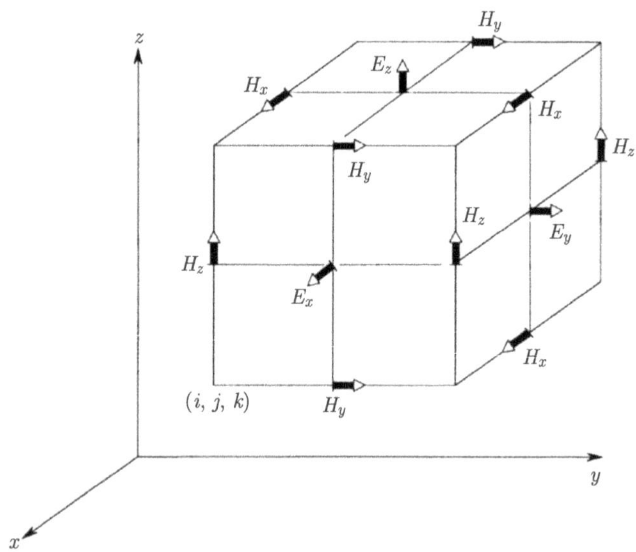

Fig. 2.2 Shows the position distribution of the components of the electric and magnetic fields in a unit grid of a three-dimensional Yee grid.

Figure 2.3 illustrates the iterative manner of the electromagnetic field evolving over time in a one-dimensional spatial grid using the FDTD algorithm. First, the electromagnetic field information is iterated by half a step to obtain the magnetic field information at half a step later. Then,

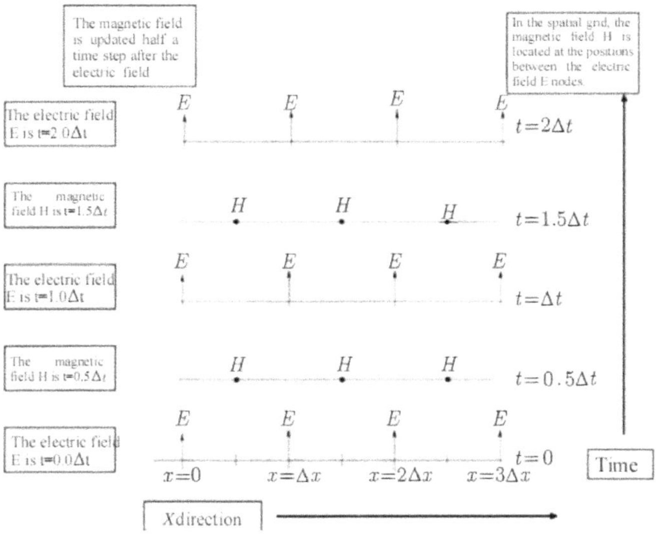

Fig. 2.3 Schematic diagram of the electromagnetic field iteration over time in the one-dimensional FDTD algorithm.

based on the updated electromagnetic field information, the electric field information at the next step is iterated by half a step, and this process is repeated.

The selection of the time step has a significant impact on the numerical stability of the FDTD solution algorithm. The time step must be less than a certain time scale; otherwise, the simulation will produce incorrect results. In PIC simulations, the time step must satisfy the Courant–Friedrichs–Lewy (CFL) condition. The underlying principle is that if a wave propagates in a discrete spatial grid and the time it takes to travel to the adjacent grid point is $\Delta x/u$, where u is the wave speed (here, the speed of light c), then the time step must resolve this time, i.e., $\Delta t \leq x/c$; otherwise, the simulation cannot capture the correct physical phenomena. In multi-dimensional cases, the time step must be less than or equal to the shortest time it takes for the wave to reach the nearest adjacent grid in the numerical simulation. This means that the size of the spatial grid imposes an upper limit on the time step. In three-dimensional simulations, the time step selection must satisfy

$$c\Delta t \leq \frac{1}{\sqrt{\frac{1}{(\Delta x)^2} + \frac{1}{(\Delta y)^2} + \frac{1}{(\Delta z)^2}}}. \tag{2.1.5}$$

This is the CFL condition for the time step. In the case of uniformly divided grids, $\Delta t \leq \frac{\Delta x}{\sqrt{3}c}$, which means that for the same grid size, three-dimensional simulations require a smaller time step than one-dimensional simulations. Current mature PIC programs generally check whether the set time step meets this condition before starting the simulation. If the set time step does not meet the CFL condition, a warning or error will be issued.

It can be seen that the setting of the time step is highly dependent on the size of the spatial grid. The size of the spatial grid needs to satisfy its ability to resolve the smallest scale of concern in the simulation. In laser wakefield acceleration simulations, this smallest scale is typically the laser wavelength. The grid size needs to be smaller than the laser wavelength, and the sampling rate within one laser wavelength affects the phase velocity of the laser propagation in the simulation. A low sampling rate can lead to severe numerical dispersion and thus incorrect simulation results. In laser ion acceleration simulations, the grid size is generally smaller and needs to resolve the laser's skin depth or the plasma's Debye length. For ultrathin nanotargets, the thickness of the target may also need to be considered. Beginners in PIC simulations are advised to refer to the grid parameter settings in published articles. Understanding the spatial grid size setting can help us better estimate the simulation speed and the occupation of computational resources. For example, in a three-dimensional simulation, if the simulation space and time remain unchanged and the spatial step is halved, considering that the time step depends on the spatial step, the computational load will increase by a factor of 16, not 8.

2.2 Introduction to PIC Simulation Software

PIC is a highly mature numerical simulation method in the field of plasma physics, which was first proposed and applied as early as 1955. Since then, scientists have developed many excellent PIC programs, some of which are open source. Generally, there is no need to write a PIC program from scratch according to the algorithm; instead, it is sufficient to learn and use existing PIC programs. If there are special requirements, improvements can be made on the basis of the original code.

Typically, PIC simulations require a significant amount of computational resources. Therefore, tasks with large computational loads need to be executed on computational clusters, which usually operate under a Unix-like operating system. To meet the needs of running such tasks, PIC programs are generally developed for Unix-like systems. Since the commonly used Windows system on personal computers has differences in the underlying aspects compared to Unix-like systems, these PIC programs cannot be directly run on the Windows system. To learn PIC programs, the most common Unix-like operating system — Linux — is introduced here. If you want to obtain a Linux runtime environment under the Windows system, you can adopt a dual-boot or virtual machine approach. Desktop Linux distributions can choose from Ubuntu, Debian, and so on, and virtual machine management software can use VMware Workstation or VirtualBox.

There are many versions of Linux, and the programs inside are usually not as directly executable as those in the Windows operating system and need to be compiled before they can be used. Fortunately, most Linux distributions come with package managers, such as the apt command in Ubuntu and Debian. Using a package manager can easily install pre-compiled software, eliminating the need for cumbersome compilation operations. However, note that the operation of the package manager requires root privileges, and as a user on a computational cluster, you typically do not have root privileges on the cluster. In this case, if the cluster does not provide the required software, you will need to compile from the source code. In addition, most PIC programs are not included in the package manager's repository and also need to be compiled manually, but the package manager can be used to install the dependency environment required by the PIC program. Therefore, learning to compile and install software in Linux is very important. The installation and compilation methods for each software are not the same, and the general installation steps are usually provided in the software's official documentation and can be consulted when needed.

Existing open-source PIC programs can be found in the English Wikipedia entry for "Particle-in-cell" (`https://en.wikipedia.org/wiki/Particle-in-cell`). Below is a specific introduction to two PIC programs, EPOCH and Smilei.

2.2.1 *EPOCH*

EPOCH is a PIC program developed in Fortran by the UK plasma physics collaborative computing project. It began in 2007 and is still being updated and maintained. Due to its design, Fortran has limited popularity in academia and a poor ecosystem compared to other languages, and it also lacks some of the advanced features of other languages. Therefore, developing new functions on top of EPOCH is quite challenging. However, as a terminal user, EPOCH is still a good choice. In addition to the basic PIC method, EPOCH also includes additional modules such as collisions, ionization, particle injection, quantum electrodynamics effects, and bremsstrahlung radiation. These additional features expand the applicability of the original PIC program, allowing it to be used to solve more physical problems in various fields. These additional modules can be selected and used as needed in the simulation. The EPOCH project code is hosted on GitHub at `https://github.com/Warwick-Plasma/epoch`, from where the latest version of the program and documentation can be downloaded.

The downloaded program is in source code form and needs to be compiled before it can be run. Compilation requires a Fortran compiler and an MPI library, which can be installed on Ubuntu using the command sudo apt install gfortran mpich. The unpacked program includes folders such as epoch1d, epoch2d, and epoch3d, corresponding to one-dimensional, two-dimensional, and three-dimensional simulation programs, respectively. The compilation can be started by running make COMPILER=gfortran in the corresponding folder. If you need to use additional modules, you can edit the Makefile to remove the comment symbol # in front of the corresponding DEFINES line and then compile. After successful compilation, the corresponding binary executable programs will be generated in the bin subfolder.

EPOCH includes almost all parameter settings in a single script file called input.deck. This means that generally, only one compilation is needed, and there is no need to recompile every time the simulation parameters are changed. In the input.deck file, parameters for the laser, particles, simulation control, output, and various additional modules can be set. For specific methods of setting parameters in the input.deck file, please refer to the official user manual. After the preparatory work is completed, the simulation can begin. Taking a two-dimensional simulation as an example, assuming that the current working directory is the epoch2d folder and the input.deck file is located in the Data subfolder, the command echo Data | ./bin/epoch2d can be run to execute the simulation. If parallel tasks need to be executed, the general MPI command (such as mpirun and mpiexec) can be used. For example, to run the task in parallel with eight processes, the command echo Data | mpirun -n 8 ./bin/epoch2d can be used.

After the simulation program is completed, a series of output data files with the suffix .sdf will be generated in the same folder as the input file input.deck, in the order of time during the simulation. This file format is a proprietary binary data format of EPOCH. The EPOCH program package includes IDL, Python, and MATLAB interfaces for reading SDF files in the SDF folder. For ease of use, it is recommended to use Python3 for data post-processing. Reading data in Python requires the installation of the NumPy library, and the Matplotlib library is often used for scientific visualization of data. The Python SDF tool also needs to be compiled and installed before use. First, enter the SDF/C folder and run the make command to compile, then enter the SDF/utilities folder and run the command ./build-3 to install the Python3 sdf module. This command will also compile two executable files, sdffilter and sdfdiff, which can be used to directly view the structure of sdf files and compare the differences between two sdf files in the command line.

After installing the Python sdf module, you can import it in a Python script file using import sdf. Then, you can use the sdf.read function to read sdf files into Python objects stored in NumPy arrays. For detailed information on the sdf.read function and the structure of the returned objects, you can view the help information by executing help(sdf) in Python. In addition, the SDF/utilities folder also includes a Python sdf helper module, which is automatically installed when the sdf module is installed. It contains some general visualization functions using Matplotlib, which are more convenient for batch processing.

2.2.2 *Smilei*

Smilei is a PIC program developed in C++ by French scientists since 2013, and it is still being updated and maintained by dedicated personnel. Smilei uses Python as the input script and the common HDF5 format for output files, with a more powerful Python post-processing module, making it more user-friendly compared to EPOCH. Additionally, Smilei uses C++ object-oriented design, which is also more friendly to developers. Like EPOCH, Smilei includes additional features such as collisions, ionization, particle injection, and quantum electrodynamics effects. Due to its object-oriented design, there is no need to enable these features in the compilation options as in EPOCH; a single compilation is sufficient for all cases. The Smilei code is also hosted on GitHub at `https://github.com/SmileiPIC/Smilei`, from where the source code can be downloaded. The usage of Smilei is similar to EPOCH: first, compile the executable file, prepare an input script file, run the program, and then perform post-processing on the output data files. There are very detailed usage instructions and explanations of various numerical algorithms on the Smilei official website (`https://smileipic.github.io/Smilei/index.html`), so they will not be repeated here. In addition, Smilei supports parallel computing with both OpenMP and MPI. For computational efficiency, it is recommended to use MPI for parallel computing (under the same number of parallel instances, the multi-threading of OpenMP is slower than the multi-process of MPI).

2.2.3 *PIC Simulation Example*

Below is an example of using the PIC simulation program Smilei to perform a complete numerical simulation of the self-focusing effect of a laser in plasma.

To perform the simulation, first prepare a script file with various input parameters (assuming that Smilei has been compiled and installed as described in the previous section). Start by setting the basic parameters of the simulation in the Main module, including the simulation type, time step, grid size, electromagnetic field boundary conditions, etc., as follows:

```
from math import pi
wavelength = 2.*pi
period = 2.*pi
Main(
    geometry = "3Dcartesian",
    timestep = 0.05*period,
    simulation_time = 72.1*period,
    cell_length = [0.1*wavelength, 0.1*wavelength, 0.1*wavelength],
    grid_length = [40.*wavelength, 20.*wavelength, 20.*wavelength],
    number_of_patches = [16, 4, 4],
    EM_boundary_conditions = [["silver-muller"], ["periodic"], ["periodic"]],
    solve_poisson = False
)
```

Next, use the LaserGaussian3D module to set a three-dimensional circularly polarized Gaussian laser pulse entering from the boundary:

```
LaserGaussian3D(
    box_side = "xmin",
    a0 = 16.5*2**0.5,
    ellipticity = 1.,
    focus = [0., Main.grid_length[1]/2., Main.grid_length[2]/2.],
    waist = 6.*wavelength,
    time_envelope = tgaussian(center=50.*period, fwhm=30.*period)
)
```

Then, set the near-critical density carbon plasma in space using the Species module:

```
Species(
    name = "electron",
    position_initialization = "regular",
    momentum_initialization = "maxwell-juettner",
    particles_per_cell = 8,
    mass = 1.0,
    charge = -1.0,
    charge_density = constant(2.4, xvacuum=3.*wavelength),
    mean_velocity = [0.0],
    temperature = [1e-5]
)

Species(
    name = "carbon",
    position_initialization = "regular",
    momentum_initialization = "maxwell-juettner",
    particles_per_cell = 1,
    mass = 12.0*1836,
    charge = 6.0,
    charge_density = constant(2.4, xvacuum=3.*wavelength),
    mean_velocity = [0.0],
    temperature = [1e-5]
)
```

Finally, set some diagnostic modules to output the required simulation results to files:

```
from numpy import s_
DiagFields(
    name = "Ez",
    every = 100,
    subgrid = s_[:, int(Main.number_of_cells[1]/2.), :],
    fields = ["Ez"]
)

DiagFields(
    name = "By_averaged",
    every = 100,
    subgrid = s_[:, int(Main.number_of_cells[1]/2.), :],
    time_average = int(2*period/Main.timestep),
    fields = ["By"]
)
DiagParticleBinning(
    name = "number_density",
    deposited_quantity = "weight",
    every = 100,
    species = ["electron"],
    axes = [
      ["x", 0., Main.grid_length[0], Main.number_of_cells[0]],
```

```
        ["z", 0., Main.grid_length[2], Main.number_of_cells[2]]
    ]
)
```

Now the simulation program can be run. Assuming the input file input.py is located in the main directory of the Smilei program, you can enter the directory and run the command ./smilei input.py to start the simulation. After the program is completed, data files such as Fields0.h5, Fields1.h5, and ParticleBinning0.h5 will be generated in the current directory, and then data visualization and other post-processing work can be carried out.

Using the Python happi module that comes with Smilei, data visualization becomes very simple. By entering the following lines of code in the command line, you can plot the data images as shown in Figures 2.4 to 2.6.

```
>>> import happi
>>> S = happi.Open('.', reference_angular_frequency_SI=1.88365e15)
>>> S.Field(0, 'Ez', timesteps=1400, units=['μm', 'fs'],
average="y":"all").plot(cmap='bwr', vsym=True)
>>> S.Field(1, 'By', timesteps=1400, units=['μm', 'fs'],
average="y":"all").plot(cmap='bwr', vsym=True)
>>> S.ParticleBinning(0, timesteps=1400, units=['μm', 'fs']).plot(cmap='hot r')
```

The above completes a complete PIC simulation process.

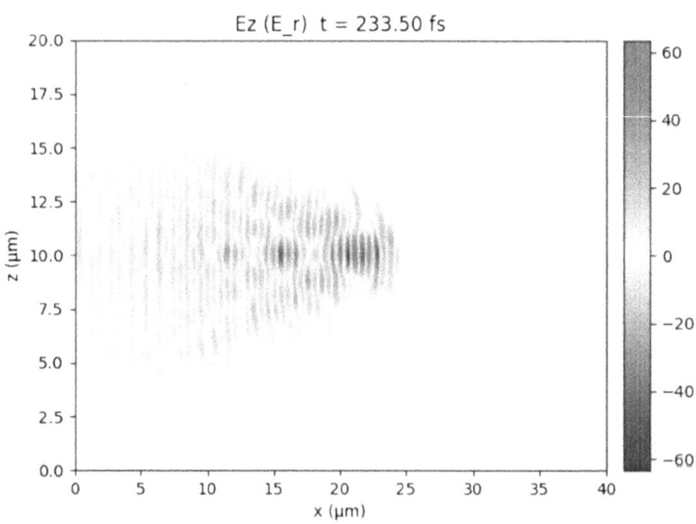

Fig. 2.4 PIC simulation: Electric field profile.

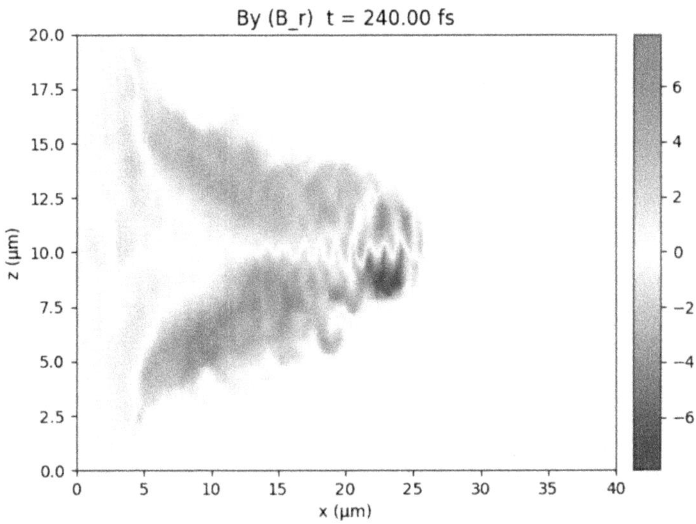

Fig. 2.5 PIC simulation: Average magnetic field profile.

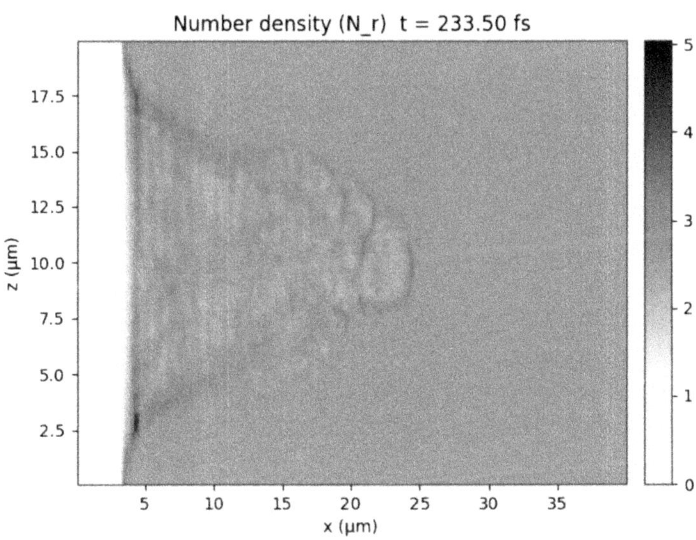

Fig. 2.6 PIC simulation: Average magnetic field profile.

2.3 Fluid Simulation

2.3.1 *Introduction to Fluid Simulation*

PIC simulations primarily focus on the kinetic evolution processes of laser-plasma interactions at the fs and μm spatiotemporal scales. When studying interactions on larger spatiotemporal scales, the computational resources required for PIC simulations exceed the acceptable range, and fluid simulations are needed. For example, in the interaction process of laser prepulses and targets, the density distribution of plasma and other macroscopic physical quantities when the main pulse arrives are usually of concern. The time interval between the prepulse and the main pulse is on the order of tens of ps, hundreds of ps, or even ns, which is much larger than the characteristic time corresponding to the electron plasma frequency ($\sim \omega_{\text{pe}}^{-1}$, on the order of fs); the expansion scale of the target under the action of the prepulse reaches the order of hundreds of μm, which is much larger than the electron Debye length (on the order of nm). On such large spatiotemporal scales, the fluid dynamics method based on continuum theory performs well. Although the fluid dynamics method will erase some physical details and lose some accuracy, its simulation error for the overall movement of plasma on large time and space scales is very small, and it can provide information on the macroscopic physical quantities of concern with the least amount of computational resources.

The fluid simulation method treats plasma as a fluid, using macroscopic average state variables — such as temperature, density, and pressure — to describe the state of plasma, focusing mainly on the overall behavior of plasma and ignoring the motion states of individual particles. The plasma fluid contains electrons and at least one type of ion, and each component must follow both general fluid dynamics laws and electrodynamics laws. Therefore, to fully characterize the plasma fluid, it is necessary to combine tens of equations, including the mass conservation, momentum conservation, and energy conservation equations of fluid dynamics, the plasma state equation, Maxwell's equations, charge conservation, and Ohm's law. Generally speaking, there is a strong coupling between electrons and ions, and under conditions where the scale is much larger than the plasma frequency (in time) and the Debye length (in space), the quasi-neutrality condition is satisfied:

$$n_{\text{e}} = \sum_{\alpha} Z_{\alpha} n_{\text{i},\alpha}, \tag{2.3.1}$$

where n_{e} is the electron density, $n_{\text{i},\alpha}$, is the density of the α-th ion species, and Z_α is the average ionization degree of the α-th ion species. Under the quasi-neutrality condition, there is no net charge density (although there can be a net current density), and the plasma fluid is electrically neutral. At this point, the system of plasma fluid equations can be greatly simplified, making it sufficient to perform simulations on existing supercomputers. Furthermore, for plasmas that exist for times much longer than the electron and ion thermal equilibrium timescales, the electron and ion components approximately satisfy the thermal equilibrium state, and the electrons and ions in the plasma can be treated as a single fluid, simplifying to a single-fluid model. Under this approximation, the fluid method for handling plasma movement becomes even more simplified, and fluid simulations can even be carried out on personal computers.

In laser-plasma interactions, plasma is generally in an extremely high-temperature state. Therefore, the heat transfer process in it needs to consider radiation. The fluid simulation method that includes the radiation transport equations is called the radiation hydrodynamics method. It solves the system of fluid dynamics equations (including the continuity equation, the motion equation, and the energy equation) and the radiation transport equations simultaneously and adds the equation of state (EOS) to close the system of equations. In laser-plasma problems, the applicable condition for the fluid dynamics method is that the plasma has entered a slowly varying state, that is, the time and space scales of the plasma state changes are much larger than the average collision time and the mean free path of plasma particles. From the perspective of the microscopic particle distribution, this means that on the scale of the fluid element, the plasma satisfies the local Maxwell distribution, and the same kind of particles or different kinds of particles have reached a thermal equilibrium state due to collisions. In this way, using macroscopic state variables (density, temperature, etc.) to describe the fluid element (much smaller than the scale of plasma change, but much larger than the particle mean free path) can both describe the local characteristics of the plasma and have statistical significance. For quasi-neutral assumptions, the expansion movement of long-time large-scale laser plasma basically meets the above conditions.

2.3.2 *Introduction to the Fluid Program MULTI*

The MULTI program is a series of radiation hydrodynamics calculation programs developed by R. Ramis *et al.* The MULTI program first appeared in 1985 [16], initially as a one-dimensional model simulation program. Later, its two-dimensional version MULTI-2D [17] and three-dimensional version MULTI-3D were developed, as well as MULTI-fs [18] suitable for femtosecond lasers.

The MULTI program is mainly used for numerical simulation research of inertial confinement fusion and related laser-matter interaction experiments. The program integrates various subprograms, such as fluid dynamics subprograms, energy deposition subprograms, and radiation transport subprograms, into a library. Users need to write their own main programs to call these subprograms in the library to solve specific problems. The more mature later versions of MULTI follow the r94 data transfer rules, so the program needs to be installed in the r94 system and written in the r94 language.

The fluid dynamics method in the MULTI program mainly adopts Lagrangian dynamics. Mass coordinates replace spatial coordinates as independent variables: $\mu(r,t) = \int_0^r \rho(r,t) 4\pi r^2 \mathrm{d}r$. For a particle moving with the fluid, μ is constant. In the absence of diffusion and reactions, the fluid components remain unchanged for a given μ value, which means that the electron specific internal energy $e^{\mathrm{e}}(\mu,t)$ and ion specific internal energy $e^{\mathrm{i}}(\mu,t)$ can be used, together with the fluid position $r(\mu,t)$ and velocity $v(\mu,t)$, as state variables. Mass density, temperature, pressure, ionization degree, and other quantities can all be expressed by these quantities. For a one-dimensional grid, the above continuous functions are discretized on the staggered grid shown in Figure 2.7. The control particles of the Lagrangian grid (with constant μ) are defined on the interfaces (referred to as interfaces) and numbered from 1 to $N+1$. The regions between interfaces (referred to as cells) are numbered from 1 to N. Cell j is between interfaces

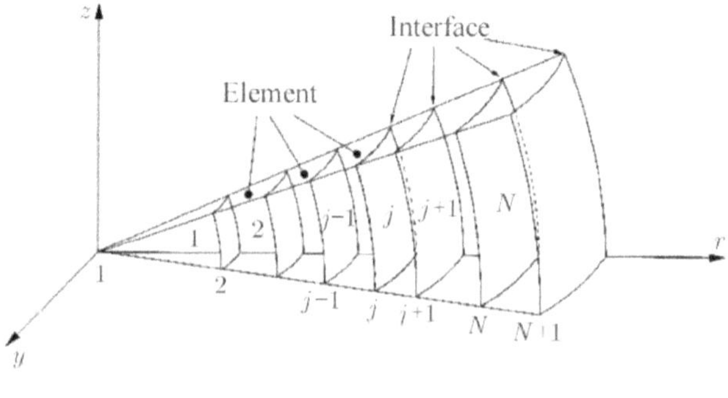

Fig. 2.7 Staggered grid (spherical). Scalars are defined on cells, and vectors are defined on interfaces.

j and $j+1$, and interface j is between cells $j-1$ and j. Generally, vectors (such as position $\boldsymbol{r}_j$ and velocity $\boldsymbol{v}_j$) are defined on the interfaces, while scalars (such as electron and ion specific internal energies e_j^{e} and e_j^{i}) are defined at the cell centers. The semi-discrete state vector $X(t)$ contains a finite number of time-continuous functions, and its evolution over time is controlled by a set of ordinary differential equations $\dfrac{\mathrm{d}X}{\mathrm{d}t} = F(X,t)$.

In the plasma model, MULTI assumes that the plasma always maintains the quasi-neutrality condition, that is, $Z_{\mathrm{i}}n_{\mathrm{i}} = n_{\mathrm{e}}$, and does not consider net current density, so $Z_{\mathrm{i}}n_{\mathrm{i}}v_{\mathrm{i}} = n_{\mathrm{e}}v_{\mathrm{e}}$, and thus ions and electrons have the same velocity $v_{\mathrm{i}} = v_{\mathrm{e}} = v$. As a result, the program also does not consider the electric and magnetic fields in the plasma.

In terms of thermal conduction modeling, MULTI uses the classical Spitzer formula for electron thermal conductivity to calculate the heat transfer coefficient. Under certain irradiation conditions, this method may result in extreme temperature gradients, and the electron thermal conductivity rate has an upper limit, so it is necessary to limit the heat flux. In MULTI, the maximum heat flux is limited by the free-streaming limit value, and an empirical coefficient is used for correction. For radiation transport, the frequency and direction grouping method is used for discretization.

In the program, laser energy deposition is treated using the quasi-static approximation: the laser is modeled as many rays, and the energy deposition of the laser rays is tracked, which is known as the ray tracing method. The law of refraction in layered media can be written as $n\sin\alpha = \text{constant}$, where n is the refractive index and α is the angle between the direction of laser propagation and the gradient direction of the refractive index. If we use mechanics as an analogy, then the law of refraction is the same as the expression for the conservation of lateral momentum of a "fictitious particle" with unit mass: the particle moves along the direction of laser propagation with a "velocity" n and is subject to a force in the direction of the density gradient, with the angle between the force direction and the "velocity" direction being α. Therefore, in the lateral direction, "momentum" is conserved, that is, $n\sin\alpha = \text{constant}$. Since the magnitude of the "velocity" only depends on n, it can be said that the "force" acting on the particle originates from the "potential energy" $-n^2/2$, and the "equation of motion" for the "particle" has the form

$$\frac{\mathrm{d}^2\boldsymbol{r}}{\mathrm{d}\xi^2} = \frac{1}{2}\nabla n^2, \tag{2.3.2}$$

where ξ is the fictitious time. From this, the differential equation for the ray trajectory and the carried power can be obtained by integrating the equation from the plasma boundary, where $|\mathrm{d}\boldsymbol{r}/\mathrm{d}\xi| = 1$. For an ideal plasma, the dispersion relation for an electromagnetic wave with angular frequency ω and wave number k satisfies $c^2k^2 = \omega^2 - \omega_{\mathrm{pe}}^2$, where $\omega_{\mathrm{pe}} = \sqrt{e^2 n_{\mathrm{e}}/\varepsilon_0 m_{\mathrm{e}}}$ is the plasma frequency and e is the electron charge. The refractive index is $n = \dfrac{ck}{\omega} = \sqrt{1 - \omega_{\mathrm{pe}}^2/\omega^2}$. The attenuation of the power I along the ray can be written as

$$\frac{\mathrm{d}I}{\mathrm{d}l} = -\frac{\nu^{\mathrm{e}}}{c}\frac{\omega_{\mathrm{pe}}^2}{\omega^2}\frac{1}{\sqrt{1-\dfrac{\omega_{\mathrm{pe}}^2}{\omega}}} = -\frac{\nu^{\mathrm{e}}}{c}\frac{1-n^2}{n}I, \tag{2.3.3}$$

where ν^{e} is the electron collision frequency. Solving the above equation yields the differential equation for the ray trajectory and the carried power.

The material properties (equation of state, ionization degree, opacity, etc.) are obtained from external tables. Overall, the physical models in the MULTI program are non-relativistic and are only applicable to situations with moderate to low light intensity. The MULTI algorithm uses the operator splitting method, which, although only first-order accurate, is very convenient and flexible, compensating for the inaccuracy of the algorithm.

The MULTI program includes one-dimensional and two-dimensional simulation programs.

The one-dimensional simulation program initially started with MULTI-1D and was upgraded to MULTI-fs for the simulation of one-dimensional

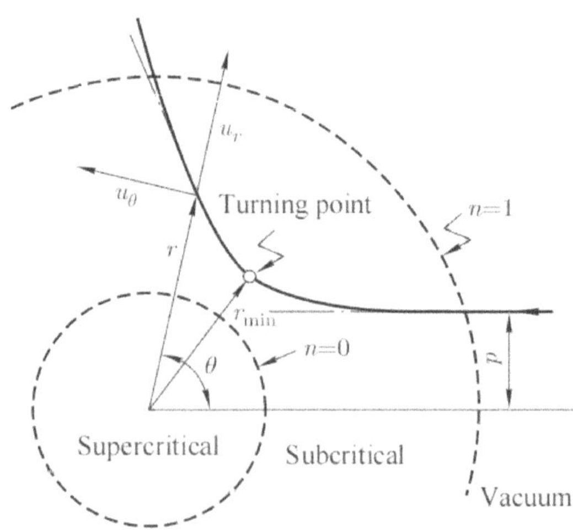

Fig. 2.8 Ray trajectory diagram.

femtosecond laser-matter interactions. The time scale of femtosecond laser-plasma interactions does not satisfy the collision condition, so the mechanisms of ray tracing and inverse bremsstrahlung absorption introduced earlier cannot be applied to femtosecond laser cases. To address this issue, MULTI-fs no longer assumes rays but directly solves Maxwell's equations. Currently, the one-dimensional program has been upgraded to the latest MULTI-IFE, which can be used for one-dimensional simulations of inertial confinement fusion.

MULTI-IFE [19] was developed in 2015, integrating the model features of MULTI-1D (2009 version) and MULTI-fs (2012 version). It is a specialized numerical tool for inertial fusion microcapsule research. It includes the relevant physics for implosion, thermonuclear ignition, and burn: two-component plasma (ions and electrons) fluid dynamics, three-dimensional laser ray tracing, thermal diffusion, multi-group radiation transport, deuterium-tritium burn, and alpha particle diffusion. The corresponding differential equations are discretized in one-dimensional Lagrangian spherical coordinates.

The two-dimensional simulation program is MULTI-2D. MULTI-2D (2009 version) uses a cylindrical coordinate system and is suitable for axisymmetric problems, mainly simulating the physical processes in the R-Z plane. It uses a single-fluid equation set, does not distinguish between electrons and ions, and has only one material temperature and one pressure. Its grid is a Lagrangian grid, which is prone to grid distortion under large deformation, leading to inaccurate or even uncomputable results. To address this issue, the arbitrary Lagrangian–Eulerian method can be used, projecting the results calculated by the Lagrangian method onto an Eulerian grid to prevent inaccurate or uncomputable results due to large grid deformation. However, the projection method in this version of the program only has first-order accuracy.

Below is a complete example of using MULTI-IFE to perform a one-dimensional simulation of a dual-layer target structure irradiated by a 300 ps (FWHM) pulse (intensity 3×10^{14} W/cm^2, wavelength 0.44 μm) on a gold foil (0.2 μm) and an aluminum foil (2 μm).

A complete simulation example includes three files: data.in, FILELIST, and the RUN file. Users mainly modify the data.in file. First, create a folder for this example in the case folder, such as "2015CPC", and then copy the FILELIST and RUN files to this folder. Next, write a data.in file containing various input parameters. This file is similar to a namelist, recording the "name - value" pairs of parameters in different blocks. First, set the basic simulation parameters in the parameters block, including general simulation settings, boundary conditions (radiation boundary conditions), electron thermal conduction model, time step, algorithm parameters, output result settings, etc., as follows:

```
&parameters
!-problem definition (general)
    igeo = 1,              ! Defines the geometry type as 1 for planar
    xmin = 0.,             ! The coordinate of the initial left boundary
    texit = 610.e-12,      ! Simulation duration
    nfuel = 0,             ! Number of DT fuel cells (0 indicates no fusion reaction)
!-problem definition (boundary conditions)
    iright = 0,            ! Free right boundary
    ileft = 0,             ! Free left boundary
    alphal = 0.,           ! Reflected radiation fraction at the left boundary
    alphar = 0.,           ! Reflected radiation fraction at the right boundary
    betal = 0.,            ! Reflected coupling fraction at the left boundary
    betar = 0.,            ! Reflected coupling fraction at the right boundary
    itype = 2,             ! Radiation time spectrum type (2 indicates constant)
    tau = 0.,              ! Radiation pulse width
    trad = 0.,             ! Radiation reference temperature
!-problem definition (modeling options)
    iradia = 1,            ! Enable radiation module
    ihydro = 1,            ! Enable fluid module
    iheation = 0,          ! Disable ion thermal conduction
    model = 0,             ! Electron collision model (0 indicates classical model)
    fheat = 8.0,           ! Model coefficient (if model==1)
    fei = 2.0,             ! Model coefficient (if model==1)
    flaser = 26.0,         ! Model coefficient (if model==1)
    zmin = 0.1,            ! Minimum ionization degree
    flf = 1e6,             ! Heat flux limiting coefficient (1e6 means no limit)
    inegpre = 0,           ! Do not allow negative pressure
!-numerical options
    nsplit = 5,            ! Number of sub-cycles
    dtmin = 0,             ! Minimum time step
    dtmax = 4e-12,         ! Maximum time step
    dtinit = 4e-12,        ! Initial time step
    dtrvar = 0.9,          ! Density-induced time step correction factor
    dttevar = 1e12,        ! Electron temperature-induced time step correction factor
    dttivar = 1e12,        ! Ion temperature-induced time step correction factor
    dtbreak = 2.,          ! Cycle break coefficient
    dtfactor = 2.,         ! Time step increase factor
    nexit = 10000,         ! Maximum number of cycles
    iwctrl = 1,            ! Energy deposition optimization coefficient
!-output control
    dt_aout = 300e-12,     ! Time interval for ASCII file output
    ns_aout = 1000000,     ! Step interval for ASCII file output
    dt_bout = 1000000.,    ! Time interval for binary file output
    ns_bout = 1,           ! Step interval for binary file output
```

```
    irad_left = 1,          ! Spectrum at this interface on the left
    irad_right = 1000,      ! Spectrum at this interface on the right
    nreduce = 1,            ! Display all grids/interfaces (do not merge grids)
/
```

Next, set the material, thickness, initial conditions (density and temperature) of the target, and divide the grid (number, grid division coefficient). The material needs to be defined through the material module, including: name, atomic number, atomic mass number, equation of state (EOS), opacity and ionization degree, radiation coefficient. The specific material data table is recorded in the program and can be referenced by id (recorded in the program manual) and file name (including relative path), linking to the current material.

```
&layer ! Aluminum foil
    nc = 20,             ! Total number of grids
    thick = 2.0e-4,      ! Thickness 2 μm
    r0 = 2.7,            ! Density 2.7 g/cm^3
    te0 = 0.03,          ! Initial electron temperature (eV)
    ti0 = 0.03,          ! Initial ion temperature (eV)
    zonpar = 0.975,      ! Grid division coefficient = Next grid size / Previous grid size
    material = 'Al',     ! Material aluminum
/
&layer ! Gold foil
    nc = 75,
    thick = 0.2e-4,
    r0 = 19.3,
    te0 = 0.03,
    ti0 = 0.03,
    zonpar = 0.975,
    material = 'Au',     ! Material gold
/
```

Finally, set the laser parameters and laser simulation method: Since this example uses a 300 ps long laser, the plasma will expand to a larger scale, and the WKB approximation method can be used to handle laser deposition.

```
&pulse wkb
    inter = 1,            ! 1 indicates laser entering from the right
    pimax = 3.0e21,       ! Maximum intensity (in CGS units)
    pitime = 300e-12,     ! Pulse width half-maximum full width 300 ps
    wl = 0.44e-4,         ! Laser wavelength
    itype = 1,            ! Laser time shape: 1 indicates sin**2, 2 indicates constant,
                          ! 4 indicates table input
    delta = 1,            ! Absorption share of the laser at the critical surface
/
```

After the data.in file is written, the simulation program can be run. In the MULTI-IFE main folder, enter

```
>> ./MULTI
```

to open the MULTI-IFE main interface. In the interface, select the created "2015CPC" folder and click "Case" → "Run" to start the simulation. After the calculation is completed, the image obtained after data visualization processing is shown in Figure 2.9. The two curves represent the mass density distribution of the dual-layer target at 100 ps and 500 ps, respectively. Here, the laser enters from the right side of the simulation area. It can be seen that at 100 ps, a longer sub-critical density region is formed in front of the target due to laser ablation, while the back of the target has not yet been severely disturbed. At 500 ps, both the front and back of the target have undergone severe expansion, and the target as a whole has moved backward. The units of the horizontal and vertical coordinates in the figure are cm and g/cm^3 in the CGS system, respectively. If other units and annotations are needed, a post-processing script needs to be written to process the obtained binary file data. Each variable in the binary file data is stored in a single line order according to the grid, and the specific rules can be consulted in the MULTI manual, which will not be repeated here.

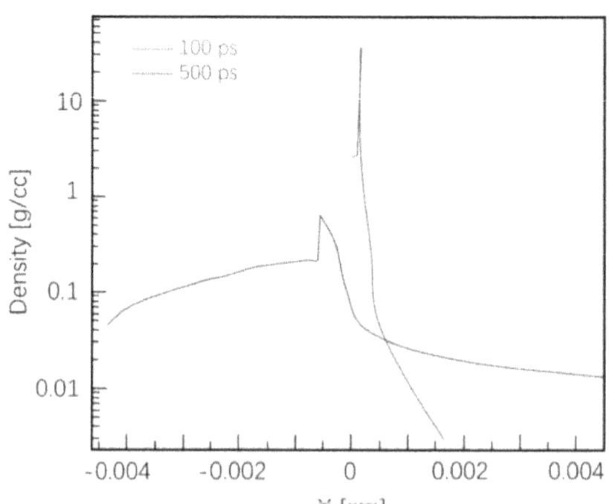

Fig. 2.9 MULTI-IFE simulation diagram: Density distribution curves at 100 ps and 500 ps. The vertical coordinate is density, and the horizontal coordinate is position, where the sharper one corresponds to 100 ps.

References

[1] Boris, J. P. Relativistic plasma simulation-optimization of a hybrid code. [J] *Proceedings of Fourth Conference on Numerical Simulations of Plasmas*, 1970: 3–67.

[2] Birdsall, C. K., and Langdon, A. B. Plasma Physics via Computer Simulation. Bristol: IOP Publishing, 1991.

[3] Hockney, R. W., and Eastwood, J. W. Computer Simulation Using Particles. London: Taylor and Francis, 1988.

[4] Chen, M., Sheng, Z., Zheng, J., *et al.* Development and application of multi-dimensional particle-in-cell codes for investigation of laser plasma interactions. *Chinese Journal of Computational Physics*, 2008, 25: 43–50.

[5] Kane, Y. Numerical solution of initial boundary value problems involving Maxwell's equations in isotropic media. *IEEE Transactions on Antennas and Propagation*, 1966, 14(3): 302–307.

[6] Taflove, A. Application of the finite-difference time-domain method to sinusoidal steady-state electromagnetic-penetration problems. *IEEE Transactions on Electromagnetic Compatibility*, 1980, EMC-22(3): 191–202.

[7] Derouillat, J., Beck, A., Pérez, F., *et al.* Smilei: a collaborative, open-source, multi-purpose particle-in-cell code for plasma simulation. *Computer Physics Communications*, 2018, 222(2018): 351–373.

[8] Buneman, O. J. P. Dissipation of currents in ionized media. *Physical Review*, 1959, 115(3): 503–517.

[9] Dawson, J. One-dimensional plasma model. *Physics of Fluids*, 1962, 5(4): 445–459.

[10] Pukhov, A. Three-dimensional electromagnetic relativistic particle-in-cell code VLPL (Virtual Laser Plasma Lab). *Journal of Plasma Physics*, 1999, 61(3): 425–433.

[11] Fonseca, R. Lecture Notes in Computer Science. Heidelberg: Springer, 2002.

[12] Nieter, C., and Cary, J. R. VORPAL: a versatile plasma simulation code. *Journal of Computational Physics*, 2004, 196(2): 448–473.

[13] Chen, M., Sheng, Z., Zheng, J., *et al.* Development and application of multi-dimensional particle-in-cell codes for investigation of laser plasma interactions. *Computational Physics*, 2008, 25(1): 43–50.

[14] Pfund, R. E. W., Lichters, R., and Meyer-Ter-Vehn, J. LPIC++, a parallel one-dimensional relativistic electromagnetic particle-in-cell code for simulating laser-plasma-interaction. AIP Conference Proceedings. *American Institute of Physics*, 1998, 426(1): 141–146.

[15] Ridgers, C. P., Brady, C. S., Duclous, R., *et al.* Dense electron-positron plasmas and bursts of gamma-rays from laser-generated quantum electrodynamic plasmas. *Physics of Plasmas*, 2013, 20(5): 056701.

[16] Ramis, R., Schmalz, R., and Meyer-Ter-Vehn, J. MULTI - a computer code for one-dimensional multigroup radiation hydrodynamics. *Computer Physics Communications*, 1988, 49(3): 475–505.

[17] Ramis, R., Meyer-Ter-Vehn, J., and Ramírez, J. MULTI 2D - a computer code for two-dimensional radiation hydrodynamics. *Computer Physics Communications*, 2009, 180(6): 977–994.
[18] Ramis, R., Eidmann, K., Meyer-Ter-Vehn, J., *et al.* MULTI-fs - a computer code for laser plasma interaction in the femtosecond regime. *Computer Physics Communications*, 2012, 183(3): 637–655.
[19] Ramis, R., and Meyer-Ter-Vehn, J. MULTI-IFE - a one-dimensional computer code for Inertial Fusion Energy (IFE) target simulations. *Computer Physics Communications*, 2016, 203(2016): 226–237.

Chapter 3

Theoretical Foundations of Laser-Plasma Interactions

3.1 Laser-Particle Interactions

This section starts from the relativistic equations of motion for a single electron, theoretically presents some physical images and fundamental laws in the interaction process between a plane electromagnetic wave and a single electron, and demonstrates typical electron motion scenarios through numerical calculations.

3.1.1 *Basic Equations of Interaction*

The single-particle theory is the foundation of laser-plasma physics. Starting from the motion equation of a single free electron in a laser field, it helps to understand the complex interactions between laser and plasma. Many studies on nonlinear laser-matter interactions begin with the trajectory of a single electron in a strong plane electromagnetic wave. This problem can be precisely solved in either the laboratory coordinate system or the average rest frame of the electron. This subsection provides the derivation process in both coordinate systems and introduces the corresponding notations. In this chapter, all derivations are carried out in the International System of Units (SI).

The momentum and energy equations for a charged particle in an electromagnetic field are given by:

$$\begin{cases} \dfrac{\mathrm{d}\boldsymbol{p}}{\mathrm{d}t} = -e(\boldsymbol{E} + \boldsymbol{v} \times \boldsymbol{B}), \\ \dfrac{\mathrm{d}(\gamma m_{\mathrm{e}} c^2)}{\mathrm{d}t} = -e\boldsymbol{v} \cdot \boldsymbol{E}, \end{cases} \tag{3.1.1}$$

where $\boldsymbol{p} = \gamma m_{\mathrm{e}} \boldsymbol{v}$ is the electron's momentum, $\boldsymbol{v}$ is the electron's velocity, γ is the relativistic factor $(1 - v^2/c^2)^{-1/2}$, e and m_{e} are the electron's charge and mass, respectively. The electric field $\boldsymbol{E}$ and magnetic field $\boldsymbol{B}$ can also be expressed in terms of the vector potential $\boldsymbol{A}$ and scalar potential Φ:

$$\begin{cases} \boldsymbol{B} = \nabla \times \boldsymbol{A}, \\ \boldsymbol{E} = -\nabla\Phi - \dfrac{\partial \boldsymbol{A}}{\partial t}. \end{cases} \tag{3.1.2}$$

In the description of laser-plasma physics, it is convenient and concise to nondimensionalize the physical quantities. For example, velocity is normalized by the speed of light in vacuum c, and time and spatial coordinates are normalized by ω_0^{-1} and k_0^{-1}, respectively, where ω_0 and k_0 are the laser frequency and wavenumber. Momentum, vector potential, and scalar potential can also be normalized as $\boldsymbol{p} = \boldsymbol{P}/m_{\mathrm{e}}c$, $\boldsymbol{a} = e\boldsymbol{A}/m_{\mathrm{e}}c$, $\phi = e\Phi/m_{\mathrm{e}}c^2$.

Considering the simple case in Eq. (3.1.1) where $|\boldsymbol{v}| \ll c$ and neglecting the $\boldsymbol{v} \times \boldsymbol{B}$ term, the linear solution is obtained as $\boldsymbol{v} = -\dfrac{\mathrm{i}e}{m_{\mathrm{e}}\omega_0}\boldsymbol{E}$, with the velocity amplitude $v_0 = eE_0/m_{\mathrm{e}}\omega_0$. The normalized vector potential amplitude a_0 is commonly used to represent the laser electric field strength:

$$a_0 = \frac{eA_0}{m_{\mathrm{e}}c} = \frac{eE_0}{m_{\mathrm{e}}c\omega_0}. \tag{3.1.3}$$

For a laser with a given wavelength λ and intensity I, there is a commonly used conversion relationship:

$$a_0^2 = I[\mathrm{W/cm^2}]\lambda^2[\mu\mathrm{m}^2]/\zeta(1.37 \times 10^{18}), \tag{3.1.4}$$

where ζ is a constant ($\zeta = 1$ for linear polarization, $\zeta = 2$ for circular polarization). Lasers with $a_0 \geq 1$ are generally defined as relativistic lasers, where relativistic effects of electrons become significant. For a 1.06 μm Nd:glass near-infrared laser, the intensity I of a relativistic laser with a_0 is approximately 10^{18} W/cm^2. After nondimensionalizing the physical quantities in Eq. (3.1.1) and expressing the electric field $\boldsymbol{E}$ and magnetic field $\boldsymbol{B}$ in terms of the vector potential $\boldsymbol{A}$ and scalar potential ϕ with $\nabla\phi = 0$, the equation can be simplified to:

$$\begin{cases} \dfrac{\mathrm{d}\boldsymbol{p}}{\mathrm{d}t} = -\left[-\dfrac{\partial \boldsymbol{a}}{\partial t} + \boldsymbol{v} \times (\nabla \times \boldsymbol{a})\right], \\ \dfrac{\mathrm{d}\gamma}{\mathrm{d}t} = \boldsymbol{v} \cdot \dfrac{\partial \boldsymbol{a}}{\partial t}. \end{cases} \tag{3.1.5}$$

To simplify the discussion, consider the interaction between a single electron and a plane wave laser propagating freely in vacuum (with the propagation direction along the positive x-axis). The vector potential of the laser field can be written as:

$$\boldsymbol{A} = (0, \delta A_0 \cos\varphi, (1-\delta^2)^{\frac{1}{2}} A_0 \sin\varphi), \tag{3.1.6}$$

where $\varphi = \omega_0 t - k_0 x + \varphi_0$, φ_0 is the initial phase. For a linearly polarized laser, $\delta = \pm 1$ or 0; for a circularly polarized laser, $\delta = \pm 1/\sqrt{2}$. The electron's momentum $\boldsymbol{p}$ is decomposed into the longitudinal component p_x (scalar) and the transverse component $\boldsymbol{p}_\perp$ (vector). The transverse and longitudinal components of the momentum in Eq. (3.1.5) are respectively given by:

$$\begin{cases} \dfrac{\mathrm{d}\boldsymbol{p}_\perp}{\mathrm{d}t} = \dfrac{\partial \boldsymbol{a}}{\partial t} + v_x \dfrac{\partial \boldsymbol{a}}{\partial x}, \\ \dfrac{\mathrm{d}p_x}{\mathrm{d}t} - \dfrac{\mathrm{d}\gamma}{\mathrm{d}t} = -v_y \left(\dfrac{\partial a_y}{\partial t} + \dfrac{\partial a_y}{\partial x} \right) - v_z \left(\dfrac{\partial a_z}{\partial t} + \dfrac{\partial a_z}{\partial x} \right). \end{cases} \tag{3.1.7}$$

For the transverse motion part (the first equality) in Eq. (3.1.7), integrating both sides yields:

$$\boldsymbol{a} - \boldsymbol{p}_\perp = \boldsymbol{p}_{\perp 0}. \tag{3.1.8}$$

For the longitudinal motion part (the second equality) in Eq. (3.1.7), since the electromagnetic wave a_y, a_z are functions of $(t - x)$ only, the right-hand side of the equation is zero. Further integrating the equation gives:

$$\gamma - p_x = \alpha. \tag{3.1.9}$$

Here, α is a constant to be determined by the initial conditions. Equations (3.1.8) and (3.1.9) give two important conservation relations, which play a wide and profound role in the analysis and treatment of the problem.

For example, consider an electron with zero initial transverse momentum $\boldsymbol{p}_{\perp 0} = 0$, then $\boldsymbol{a} - \boldsymbol{p}_\perp = 0$ and $\gamma - p_x = \alpha$. According to $\gamma = (1 + p_x^2 + p_\perp^2)^{1/2}$, by eliminating γ, we can obtain the relationship between the electron's transverse and longitudinal momenta:

$$p_x = \frac{1 - \alpha^2 + p_\perp^2}{2\alpha}. \tag{3.1.10}$$

This relation represents the general solution of the free electron's motion in the electromagnetic field. Additionally, in the case of an initially stationary electron, i.e., $\boldsymbol{p}_0 = 0$, $\gamma_0 = 1$, it is easy to derive $\alpha = 1$, $\boldsymbol{p}_\perp = \boldsymbol{a}$, $p_x = a^2/2$, $\gamma - 1 = p_x = a^2/2$, that is, the electron's longitudinal momentum is proportional to the first power of the laser field strength, while its kinetic energy and transverse momentum are proportional to the second power of the laser field strength.

When the laser's normalized field amplitude $a_0 \ll 1$ (non-relativistic laser), the changes in γ and p_x are of the second order of magnitude, so the electron's main motion is transverse. In the case of linearly polarized laser, the electron oscillates parallel to the direction of the magnetic vector potential at the laser frequency, with its maximum oscillation velocity being $v_{\max} = eE_0/\gamma m_e\omega_0 = a_0c/\gamma$. If it is circularly polarized light, the electron will continuously rotate along the direction of the magnetic vector potential. However, when $a_0 \sim 1$ (relativistic laser) or $a_0 \gg 1$ (ultra-relativistic laser), the electron's longitudinal motion becomes significant, even surpassing the transverse motion. As shown in Figure 3.1, the upper two figures depict the electron trajectories in the interaction of linearly polarized and circularly polarized lasers with a single electron when $a_0 \ll 1$, where the electron mainly undergoes transverse motion; the lower two figures show the case when $a_0 > 1$, at which point the electron's longitudinal motion becomes noticeable. This is because the electron's transverse oscillation velocity approaches the speed of light in the laser electric field, making the longitudinal Lorentz force comparable to the transverse laser electric field force. Moreover, according to the above derivation, the electron's energy depends on the laser electric field strength it experiences. After the pulse passes, the electron experiences a normalized laser electric field strength of zero, $a_0 = 0$. An initially stationary electron remains stationary and does

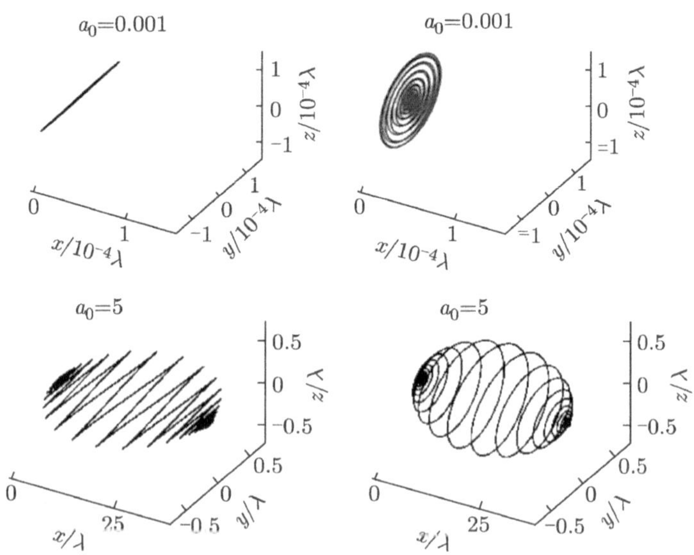

Fig. 3.1 Trajectories of initially stationary electrons in linearly polarized (upper left) and circularly polarized (upper right) laser pulses with $a_0 \ll 1$, and in linearly polarized (lower left) and circularly polarized (lower right) laser pulses with $a_0 > 1$.

not gain energy from the plane wave laser pulse. Therefore, a pure plane wave laser pulse interacting with a single electron cannot directly accelerate the electron. The electron only acquires a temporary energy during the rising edge of the laser and returns the energy to the laser during the falling edge.

The momentum and trajectory expressions of an initially stationary electron in the laboratory coordinate system under the action of a laser field are given by Eqs. (3.1.11) and (3.1.12):

$$\begin{cases} p_x = \dfrac{a_0^2}{4}[1 + (2\delta^2 - 1)\cos 2\varphi], \\ p_y = \delta a_0 \cos\varphi, \\ p_z = (1 - \delta^2)^{1/2} a_0 \sin\varphi, \end{cases} \tag{3.1.11}$$

$$\begin{cases} x = \dfrac{1}{4}a_0^2\left(\varphi + \dfrac{2\delta^2 - 1}{2}\sin 2\varphi\right), \\ y = \delta a_0 \sin\varphi \\ z = -(1 - \delta^2)^{1/2} a_0 \cos\varphi. \end{cases} \tag{3.1.12}$$

The electron trajectory diagram is shown in Figure 3.2 (note that the horizontal coordinate is normalized; for weak light intensity, the electron's longitudinal drift is very small).

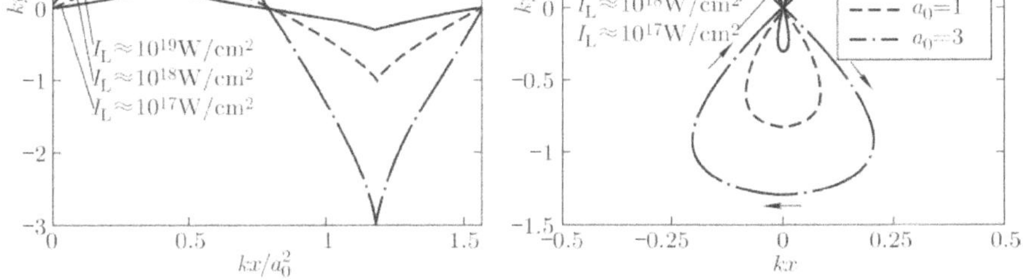

Fig. 3.2 Electron trajectory diagram [1].

From the longitudinal drift motion of the electron in the laboratory coordinate system, it can be seen that by choosing an appropriate constant α, the electron's average drift velocity can be made zero. Next, the electron's motion will be analyzed in the electron's average rest frame. Setting $\overline{p}_x = 0$ in Eq. (3.1.10) yields:

$$1 - \alpha^2 + \overline{p_\perp^2} = 1 - \alpha^2 + \overline{a^2} = 0. \tag{3.1.13}$$

Taking the average over one laser period, the constant α can be determined as $\alpha = \left(1 + \dfrac{a_0^2}{2}\right)^{1/2}$, and then the momentum and trajectory equations in the electron's average rest frame are obtained:

$$\begin{cases} p_x = (2\delta^2 - 1)\dfrac{a_0^2}{4\alpha}\cos 2\varphi, \\ p_y = \delta a_0 \cos\varphi, \\ p_z = (1-\delta^2)^{\frac{1}{2}} a_0 \sin\varphi, \end{cases} \tag{3.1.14}$$

$$\begin{cases} x = \left(\delta^2 - \dfrac{1}{2}\right) q^2 \sin 2\varphi, \\ \boldsymbol{r}_\perp = 2[\delta q \sin\varphi, -(1-\delta^2)^{\frac{1}{2}} q\cos\varphi], \end{cases} \tag{3.1.15}$$

where the parameter $q = a_0/2\alpha$. For a linearly polarized light ($\delta = 1$), the electron's trajectory in the electron's average rest frame presents an "8" shape, as shown in the right figure of Figure 3.2. When the laser intensity is very weak, the Lorentz force is much smaller than the electric field force, and the electron oscillates in the direction of the laser electric field. As the laser intensity increases, the Lorentz force strengthens, and the electron's motion in the axial direction (x direction) becomes significant, with its amplitude gradually exceeding that in the transverse direction (y direction).

3.1.2 *Laser Radiation Pressure*

Strictly speaking, the discussion in the previous subsection is only applicable to plane waves or electromagnetic waves with slowly varying amplitudes in space. However, the ultra-intense and ultra-short pulse lasers actually used are not like this. For example, a tightly focused laser beam will have a large intensity gradient change over a few wavelengths in the radial direction, and the laser intensity may also change significantly within a few laser cycles in the axial direction. At this time, the electron will also be subjected to the radiation pressure of the laser. The concept of laser radiation pressure is introduced below. To simplify the discussion, the situation of non-uniform laser intensity distribution in the transverse direction is mainly considered.

In fact, there is still some controversy in the laser-plasma community on how to strictly define radiation pressure. However, it can be intuitively understood as: radiation pressure is the gradient of the time-averaged laser intensity. The derivation of radiation pressure given here is also not entirely rigorous, but it can be used as an intuitive understanding of the influence of laser radiation pressure on particle motion. Consider the non-relativistic case: an electron oscillates at the center of a focused laser beam. Similarly, the laser is still chosen to propagate along the positive x-axis, but at this time the laser electric field strength varies with the radial position $\boldsymbol{r}$. The motion equation of the electron in the transverse y direction can be written as:

$$\frac{\partial v_y}{\partial t} = -\frac{e}{m_\mathrm{e}} E_y(\boldsymbol{r}). \tag{3.1.16}$$

Performing a Taylor expansion of $E_y(\boldsymbol{r})$ in the y direction yields:

$$E_y(r) = E_0(y)\cos\phi + y\frac{\partial E_0(y)}{\partial y}\cos\phi + \cdots, \tag{3.1.17}$$

where $\phi = \omega_0 t - k_0 x$. Taking only the zeroth-order term $E_0(y)\cos\phi$ of $E_y(r)$ and substituting it into Eq. (3.1.16), the zeroth-order expressions for the particle velocity and displacement can be obtained as $v_y^{(0)} = -v_\mathrm{os}\sin\phi$; $y^{(0)} = v_\mathrm{os}\cos\phi/\omega$, where $v_\mathrm{os} = eE_0/m\omega_0$, which is the electron oscillation motion caused by the zeroth-order term of the laser electric field. Taking the zeroth-order and first-order terms of $E_y(r)$, and combining $v_y^{(0)}$ and $y^{(0)}$ into Eq. (3.1.16), we get:

$$\frac{\partial v_y^{(1)}}{\partial t} = -\frac{e^2}{m_\mathrm{e}^2\omega_0^2} E_0 \frac{\partial E_0(y)}{\partial y}\cos^2\phi. \tag{3.1.18}$$

It can be seen that the part $v_y^{(1)}$ of the electron motion originates from the gradient of the transverse variation of the laser electric field. Multiplying the above equation by the particle mass m_e and taking the cycle average yields the expression for radiation pressure:

$$f_\mathrm{p} \equiv m_\mathrm{e}\overline{\frac{\partial v_y^{(2)}}{\partial t}} = -\frac{e^2}{4m_\mathrm{e}\omega_0^2}\frac{\partial E_0^2}{\partial y}. \tag{3.1.19}$$

This force acts to push the electron from the high-intensity region of the laser to the low-intensity region, thus causing the electron to move away from the center of the focused laser beam, as shown in Figure 3.3.

Another simple way to understand it is: in the first half of the cycle, the electron moves away from the center strong field region under the action of the laser electric field. In the second half of the cycle, when the laser electric field reverses, the electron feels a weaker laser electric field. Therefore, at

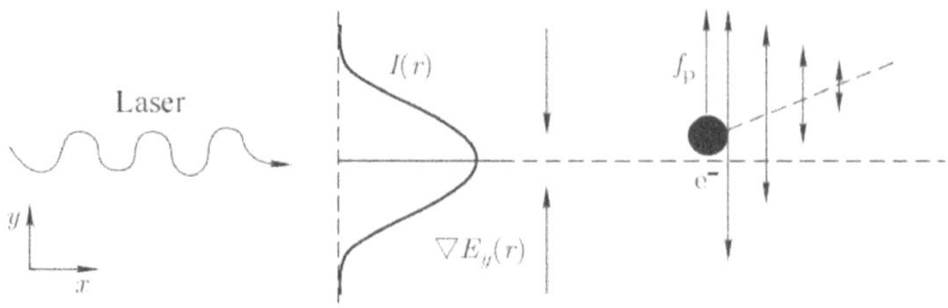

Fig. 3.3 Schematic diagram of the radial ponderomotive force of a focused laser beam. Due to the effect of the radial ponderomotive force, electrons move from the central region with high electric field strength to the peripheral region with low electric field strength [2].

the end of the second half of the cycle, it cannot return to its original position but drifts in the direction of the intensity gradient while gaining energy. It should be noted that only the "radiation pressure" experienced by the electron in the transverse direction is described here. Strictly speaking, radiation pressure is along the gradient direction of the intensity decrease and is not limited to the transverse direction. To obtain the result in this direction, it is also necessary to consider the longitudinal drift caused by the Lorentz force, that is, to consider the relativistic effects of electron motion. Starting from the Lorentz equation in Eq. (3.1.5), we obtain:

$$\frac{\partial \boldsymbol{p}}{\partial t} + (\boldsymbol{v} \cdot \nabla)\boldsymbol{p} = e\left[\frac{\partial \boldsymbol{A}}{\partial t} - \boldsymbol{v} \times (\nabla \times \boldsymbol{A})\right]. \tag{3.1.20}$$

The electron momentum is divided into fast-varying part $\boldsymbol{p}^{\mathrm{f}}$ and slow-varying part $\boldsymbol{p}^{\mathrm{s}}$ according to the time scale of motion, where the fast-varying part is directly caused by the laser electric field, so $\boldsymbol{p}^{\mathrm{f}} = e\boldsymbol{A}$, and the slow-varying part is the time-averaged effect of the electron in the laser intensity gradient. Combining the following equation:

$$\boldsymbol{v} \times (\nabla \times \boldsymbol{p}) = \frac{1}{m_{\mathrm{e}}\gamma}\boldsymbol{p} \times \nabla \times \boldsymbol{p} = \frac{1}{2m_{\mathrm{e}}\gamma}\nabla|\boldsymbol{p}|^2 - \frac{1}{m_{\mathrm{e}}\gamma}(\boldsymbol{p} \cdot \nabla)\boldsymbol{p}, \tag{3.1.21}$$

taking the average over one laser cycle yields the relativistic radiation pressure:

$$\boldsymbol{f}_{\mathrm{p}} = \frac{\mathrm{d}\boldsymbol{p}^{\mathrm{s}}}{\mathrm{d}t} = -m_{\mathrm{e}}c^2\nabla\overline{\gamma}, \tag{3.1.22}$$

where $\overline{\gamma} = (1 + p_{\mathrm{s}}^2/m_{\mathrm{e}}^2c^2 + a_0^2/2)^{1/2}$. If $a_0 \ll 1$, the above equation can be reduced to Eq. (3.1.19). The above derivation is not entirely rigorous, but this conclusion is consistent with the results obtained by more complex methods such as the Lagrangian method.

It should be noted that the radiation pressure expression derived in the single-particle model can also be used in the plasma situation. In the actual plasma, there are no single free electrons. Generally, collective effects such as electron plasma waves need to be considered, and the fluid model is usually used to describe them. The radiation pressure expression derived in the fluid model is completely consistent with Eq. (3.1.22).

3.2 Laser-Plasma Interactions

3.2.1 *Two-Fluid Description of Plasma*

The previous section discussed the interaction process between a laser and a single electron, while the main research object of this book is plasma. Plasma, as the fourth state of matter, is a macroscopically quasi-neutral state of matter composed of a large number of electrons, ions, and some neutral particles, with collective effects as the main feature. Under the action of a laser, atoms are ionized to produce a large number of unbound electrons, which move in the background of positively charged ions. Due to the Coulomb interaction between charges, charge screening occurs, and the Coulomb screening potential of charged particles is expressed as:

$$\varphi = \frac{q}{4\pi\varepsilon_0 r}\mathrm{e}^{-\frac{r}{\lambda_{\mathrm{D}}}}, \tag{3.2.1}$$

where $\lambda_{\mathrm{D}} = \sqrt{\varepsilon_0 k_{\mathrm{B}} T_{\mathrm{e}}/n_{\mathrm{e}} e^2}$ is the Debye length, k_{B} is the Boltzmann constant, T_{e} is the electron temperature of the plasma, and n_{e} is the electron number density. The Debye length characterizes the minimum scale of charged particles forming plasma, within which there are a large number of charged particles, and it appears quasi-neutral to the outside. When plasma is disturbed by external perturbations, it will exhibit collective oscillation behavior. Since the mass of electrons is much smaller than that of various ions, ions are often treated as a fixed positively charged background in the treatment process. Under this ion background, when electrons are disturbed, they will be pulled back due to the Coulomb attraction of ions. Considering a simple one-dimensional model, from the divergence equation of the electric field $\nabla \cdot \boldsymbol{E} = n_{\mathrm{e}}/\varepsilon_0$, the electric field can be approximately given as $E = n_{\mathrm{e}} e x/\varepsilon_0$, and the electron motion equation is written as:

$$m_{\mathrm{e}}\frac{\mathrm{d}^2 x}{\mathrm{d}t^2} = -eE = -e\frac{n_{\mathrm{e}} e}{\varepsilon_0}x. \tag{3.2.2}$$

It can be seen that electrons oscillate around ions in a simple harmonic motion, also known as plasma oscillations, with an oscillation frequency:

$$\omega_{\mathrm{pe}} = \sqrt{\frac{n_{\mathrm{e}} e^2}{m_{\mathrm{e}} \varepsilon_0}}. \tag{3.2.3}$$

Below, the two-fluid model of plasma is introduced, where electrons are treated as one fluid and ions as another. In plasma without a large external magnetic field, there exist high-frequency electron plasma waves and low-frequency ion acoustic waves. In 1938, Vlasov proposed the kinetic equation for collisionless plasma:

$$\frac{\partial f_j(\boldsymbol{x},\boldsymbol{v},t)}{\partial t} + \boldsymbol{v} \cdot \frac{\partial f_j(\boldsymbol{x},\boldsymbol{v},t)}{\partial \boldsymbol{x}} + \frac{e_j}{m_j}[\boldsymbol{E}(\boldsymbol{x},t) + \boldsymbol{v} \times \boldsymbol{B}(\boldsymbol{x},t)] \cdot \frac{\partial f_j(\boldsymbol{x},\boldsymbol{v},t)}{\partial \boldsymbol{v}} = 0, \tag{3.2.4}$$

where $f_j(\boldsymbol{x},\boldsymbol{v},t)$ represents the function of the position of particles of type j in the phase space $(\boldsymbol{x},\boldsymbol{v})$ varying with time. The properties of a collisionless plasma can be fully described by the Vlasov equation and Maxwell's equations.

The fluid equations of plasma are derived from the Vlasov equation by taking moments with respect to velocity. The continuity equation for particle density and the equation of motion for particles are:

$$\begin{cases} \dfrac{\partial n_j}{\partial t} + \nabla \cdot (n_j \boldsymbol{u}_j) = 0, \\ m_j n_j \dfrac{\mathrm{d}\boldsymbol{u}_j}{\mathrm{d}t} = q_j n_j (\boldsymbol{E} + \boldsymbol{u}_j \times \boldsymbol{B}) - \nabla P_j - m_j n_j \gamma_{ij} (\boldsymbol{u}_i - \boldsymbol{u}_j), \end{cases} \tag{3.2.5}$$

where γ_{ij} is the collision frequency between electrons and ions. The pressure P of the fluid is given by the equation of state for the fluid: when the characteristic frequency ω and wavenumber k of the physical process experienced by the plasma (e.g., interaction with a laser) satisfy $\omega/k \gg u_j$, the isothermal equation of state $P_j = n_j T_j$ is considered; when $\omega/k \ll u_j$, the adiabatic equation of state $P_j/n_j^{\gamma} = \text{constant}$ is considered, where $\gamma = (N+2)/N$ and N is the number of degrees of freedom; when ω/k is comparable to u_j, the details of the particle velocity distribution need to be considered, and it is necessary to return to the Vlasov equation for description. The system of Eqs. (3.2.5) together with the equation of state for the fluid constitutes the two-fluid model, and combined with the Maxwell equations in plasma:

$$\begin{cases} \nabla \cdot \boldsymbol{E} = \dfrac{1}{\varepsilon_0} \sum\limits_j n_j q_j, \\ \nabla \cdot \boldsymbol{B} = 0, \\ \nabla \times \boldsymbol{E} = -\dfrac{\partial \boldsymbol{B}}{\partial t}, \\ \nabla \times \boldsymbol{B} = \mu_0 \sum\limits_j n_j q_j \boldsymbol{u}_j + \mu_0 \varepsilon_0 \dfrac{\partial \boldsymbol{E}}{\partial t}, \end{cases} \tag{3.2.6}$$

constitutes the complete description of the two-fluid model of plasma.

First, analyze the propagation of high-frequency fields in plasma. The high-frequency part of the plasma physical quantities is regarded as the response to the high-frequency field, and the low-frequency part is considered as the contribution to the time average. Ions have a large mass and are difficult to respond to high-frequency fields, so only the contribution of electrons to the high-frequency current needs to be considered, and the electrons undergo an adiabatic process with a characteristic thermal velocity of $v_{\mathrm{te}} = \sqrt{k_{\mathrm{B}} T_{\mathrm{e}}/m_{\mathrm{e}}}$. Considering a small perturbation of the system $n_{\mathrm{e}} = n_0 + n_{\mathrm{e}1}$, $\boldsymbol{u} = \boldsymbol{u}_0 + \boldsymbol{u}_{\mathrm{e}1}$, $\boldsymbol{E} = \boldsymbol{E}_1$, where the subscript 1 indicates the first-order small quantity, and retaining only the first-order small quantity terms in the above two sets of equations, the linearized equations can be obtained:

$$-\frac{e}{m_{\mathrm{e}}} \nabla P_{\mathrm{e}} = v_{\mathrm{te}}^2 \left(\frac{\nabla n_{\mathrm{e}1}}{n_0} - \frac{1}{n_0^2} n_{\mathrm{e}1} \nabla n_0 \right). \tag{3.2.7}$$

Considering that the plasma is a response to the high-frequency field of the form $\boldsymbol{E}_1 = \boldsymbol{E}_0 \exp\{-i(k \cdot x - \omega t)\}$, and combining Poisson's equation $\nabla \cdot \boldsymbol{E}_1 = -e n_{\mathrm{e}1}/\varepsilon_0$ and the electric field propagation equation, we obtain:

$$\nabla^2 \boldsymbol{E}_1 + \frac{\omega^2}{c^2} \varepsilon(\boldsymbol{r}) \boldsymbol{E}_1 = \left[\left(1 - \frac{v_{\mathrm{te}}^2}{c^2} \right) \nabla + \frac{v_{\mathrm{te}}^2}{c^2} \frac{\nabla n_0}{n_0} \right] (\nabla \cdot \boldsymbol{E}_1), \tag{3.2.8}$$

where $\varepsilon(\boldsymbol{r}) = 1 - \omega_{\mathrm{pe}}^2/\omega^2 (1 + \mathrm{i}\gamma_{\mathrm{e,i}}/\omega)$ is the dielectric constant of the plasma, and it is usually assumed that $\gamma_{\mathrm{e,i}} \ll \omega$. For uniform plasma, substituting the transverse wave condition $\boldsymbol{k} \cdot \boldsymbol{E} = 0$ and the longitudinal wave condition $\boldsymbol{k} \times \boldsymbol{E} = 0$ into the above equation, we obtain the dispersion relations of electromagnetic waves and electrostatic plasma waves in uniform plasma, respectively:

$$\omega^2 = \omega_{\mathrm{pe}}^2 + c^2 k^2, \tag{3.2.9}$$

$$\omega^2 = \omega_{\mathrm{pe}}^2 + 3 v_{\mathrm{te}}^2 k^2. \tag{3.2.10}$$

For the low-frequency case, the motion of ions also needs to be considered. Electrons will follow the motion of ions. In the case of low-frequency long waves (i.e., the wavelength is much larger than the Debye screening distance), the thermodynamic state of electrons is given by the isothermal equation of state. Using a similar derivation, the dispersion relation of ion acoustic waves can be obtained as:

$$\omega^2 = v_s^2 k^2, \tag{3.2.11}$$

where $v_s = \sqrt{(\gamma_i T_i + \gamma_e T_e)/m_i}$ is the ion sound speed.

3.2.2 *Propagation of Laser in Plasma*

The dispersion relation of laser in plasma $\omega^2 = \omega_{pe}^2 + c^2k^2$ has been given above, and the refractive index of the laser in plasma can be derived as:

$$\eta = \frac{ck}{\omega} = \sqrt{1 - \frac{\omega_{pe}^2}{\omega^2}}. \tag{3.2.12}$$

This equation indicates that when $\omega_{pe} < \omega$, the refractive index η is less than 1, and the laser can propagate in plasma; when $\omega_{pe} > \omega$, the refractive index is imaginary, meaning that the laser intensity decays rapidly in plasma and cannot propagate within it. This critical point $\omega_{pe} = \omega$ defines the critical density of plasma. For non-relativistic lasers, the critical density can be obtained from Eq. (3.2.12) as:

$$n_c = \frac{m_e \varepsilon_0}{e^2}\omega^2 \approx 1.1 \times 10^{21} \left(\frac{\lambda}{\mu\text{m}}\right)^{-2} \text{cm}^{-3}. \tag{3.2.13}$$

For relativistic lasers with $a_0 \geq 1$, relativistic effects need to be considered. The cycle-averaged relativistic factor can be expressed as $\gamma = \sqrt{1 + a_0^2/2}$. In this case, the critical density that a relativistic laser can penetrate is γ times that of a non-relativistic laser, a phenomenon known as laser relativistic self-transparency. At this time, the refractive index of the relativistic laser in plasma is also modified to $\sqrt{1 - n_e/\gamma n_c}$. It can be seen that the refractive index of the laser in plasma increases with the increase of laser intensity. For a common Gaussian pulse, the laser intensity is the highest on the axis, so the refractive index is the largest and the phase velocity is the smallest. In this way, plasma will focus the laser like a prism (as shown in Figure 3.4), an effect known as relativistic self-focusing. At the same time, the non-uniform distribution of laser intensity in the transverse direction will cause radiation pressure to push electrons to the sides, thus forming a density distribution with low center and high sides

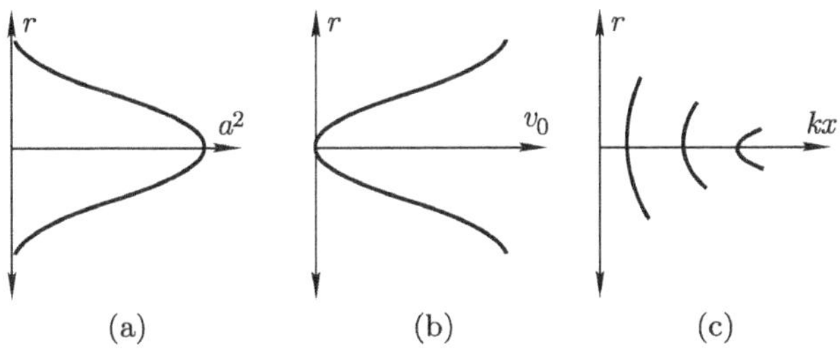

Fig. 3.4 Schematic diagram of relativistic self-focusing [3].

(the refractive index is the largest in the middle), which will also lead to the self-focusing effect of the laser, known as radiation pressure self-focusing. Theoretical results show that there is a power threshold for self-focusing of laser in plasma $P_{\mathrm{L}} \approx 16.2(\omega/\omega_{\mathrm{pe}})^2 10^9$ W, and only lasers with power greater than this can exhibit self-focusing in the corresponding plasma.

In addition, the non-uniform distribution of laser intensity in the longitudinal direction will also bring about nonlinear effects. The group velocity of the laser is greater in the region of strong intensity than in the region of weak intensity, which will cause the laser pulse to compress and the front of the pulse to steepen during propagation in plasma, an effect known as laser self-phase modulation. The self-focusing and self-phase modulation effects of the laser have been verified in both simulations and experiments. It should be pointed out that the nonlinear effects such as relativistic self-focusing and self-phase modulation are physical phenomena that occur when the plasma density is below the critical density or near the critical density. Adjusting the laser intensity and envelope, these effects can play a guiding role and are also important contents in plasma optics research. For electron density much higher than the critical density ($n_{\mathrm{e}} \gg n_{\mathrm{c}}$), such as in solid targets, the laser cannot penetrate the plasma, and the interaction range of the laser with plasma is limited to the skin depth $l_{\mathrm{s}} \approx c/\omega_{\mathrm{pe}}$.

3.3 Electron Heating (Energy Absorption) Mechanisms

Under the current laser parameters with intensity $I \leq 10^{22}$ W/cm^2, the laser cannot directly accelerate ions. The laser first transfers energy to electrons, and then the thermal pressure of electrons or the radiation pressure of the laser can maintain the separation between electrons and ions, and ions can obtain energy through the electrostatic separation field. In the process of laser-plasma interaction, electron heating is an extremely important process. At present, the main electron heating mechanisms include collisional absorption, resonant absorption, vacuum heating, $\boldsymbol{J} \times \boldsymbol{B}$ heating, and direct laser acceleration, which are introduced one by one below.

3.3.1 *Collisional Absorption*

Collisional absorption occurs in the range of relatively low laser intensity (about 10^{13} W/cm^2 to 10^{15} W/cm^2), caused by collisions between electrons and ions, also known as inverse bremsstrahlung absorption. Electrons vibrate at high frequency under the action of the laser electric field and collide with ions. The collision effect transfers the kinetic energy of electrons to ions, and the energy of the laser is thereby absorbed by the plasma. Ignoring collective and quantum effects, when $\hbar\omega \ll k_{\mathrm{B}}T_{\mathrm{e}}$, the linear absorption coefficient of the laser due to collisional effect can be obtained as:

$$K_{\mathrm{ab}} \approx \frac{Zn_{\mathrm{e}}^2}{T_{\mathrm{e}}^{3/2}(1 - n_{\mathrm{e}}/n_{\mathrm{c}})^{1/2}}, \tag{3.3.1}$$

where Z, T_{e}, n_{e}, and n_{c} are the ionization degree, electron temperature, electron density, and critical density of the plasma, respectively. From Eq. (3.3.1), it can be seen that, provided the laser can penetrate the plasma, as the plasma density increases, the collision frequency between electrons and ions also increases. Therefore, collisional absorption mainly occurs in the region near the critical density, and the hot electron temperature generated by collisional absorption is generally below the keV level. However, as the laser intensity increases, the rise in plasma temperature directly leads to a rapid decrease in the collision frequency between electrons and ions, the effect of inverse bremsstrahlung absorption weakens, and other non-collisional absorption mechanisms begin to dominate.

3.3.2 *Resonant Absorption*

Resonant absorption [8,9] is an energy conversion process from electromagnetic waves to electrostatic waves that occurs near the critical density of plasma under non-relativistic laser conditions ($\sim 10^{16}$ W/cm^2). Consider a p-polarized laser incident at an angle θ (the incident angle θ is defined as the angle between the laser wave vector direction $\boldsymbol{k}$ and the plasma density gradient direction $\widehat{\boldsymbol{z}}$) onto a non-uniform density plasma. The laser electric vector has a component parallel to the electron density gradient direction, i.e., $\boldsymbol{E} \cdot \nabla n_{\mathrm{e}} \neq 0$, so the electric field can drive electrons to oscillate along the density gradient direction, resulting in charge density fluctuations. At the plasma critical density, this density fluctuation is resonant, thereby exciting electron plasma waves.

Laser oblique incidence, p-polarization, and plasma non-uniformity are the three necessary conditions for resonant absorption to occur. Figure 3.5 shows the schematic process of a p-polarized laser being reflected and exciting electron plasma waves resonantly in a density-gradient plasma. Part of the laser energy is transferred to the plasma wave through linear or nonlinear modes, heating the plasma wave rapidly until it breaks down and produces superthermal electrons. The corresponding hot electron temperature T_e has an empirical scaling relationship with laser intensity [10]:

$$T_e \approx 14(I\lambda)^{1/3}T_b^{1/3}, \tag{3.3.2}$$

where I is the laser intensity in units of 10^{16} W/cm^2, λ is the laser wavelength in units of μm, and T_b is the background electron temperature in units of keV.

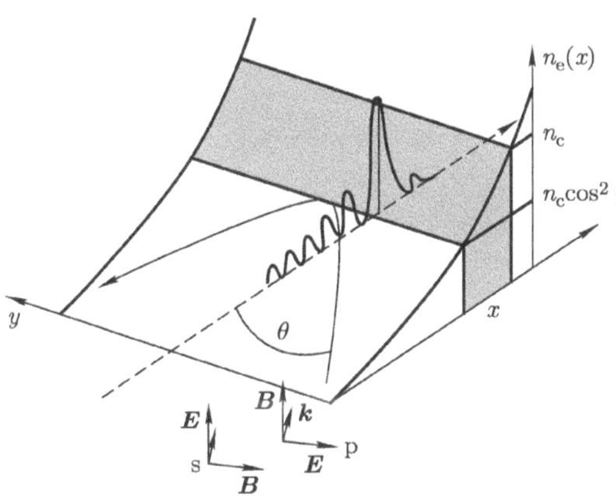

Fig. 3.5 Resonant absorption of obliquely incident p-polarized laser near the critical density surface of non-uniform plasma [2].

3.3.3 *Vacuum Heating*

Vacuum heating, proposed by Brunel [11], is also known as "Not-so-resonant, resonant absorption." As the name implies, vacuum heating is related to resonant absorption. In fact, they represent absorption mechanisms in different plasma density distribution states. When the plasma density scale length is large (density varies slowly), resonant absorption dominates, and the laser electric field can drive a strong plasma wave near the critical density surface, transferring energy to electrons. When the plasma density variation is steep, with a density scale length less than one laser wavelength, the resonance condition cannot be met, and vacuum heating dominates.

Consider a linearly polarized laser incident at an angle θ onto a plasma target whose density rises sharply from zero to much greater than the critical density. Near the critical density, the laser will resonantly drive a plasma wave with an amplitude of E_{p}. Under the steep boundary condition, E_{p} can be approximated as the laser electric field E_0. This will form a standing wave field with an amplitude of $E_{\mathrm{d}} = 2E_0 \sin\theta$ in front of the target. This oscillating electric field can violently pull electrons out from the target surface, and when the field reverses direction, the electrons will reverse and accelerate into the plasma, with an oscillation amplitude along the density gradient direction of $x_{\mathrm{p}} \approx eE_0/m_{\mathrm{e}}\omega_0^2$. When the electrons return to their initial position, their velocity should be $v_{\mathrm{d}} \approx 2v_{\mathrm{os}} \sin\theta$, where $v_{\mathrm{os}} = eE_0/m_{\mathrm{e}}\omega_0$ is the electron oscillation velocity in the laser field. At this time, the plasma density is much higher than the critical density, and the laser cannot penetrate the plasma, so the electrons leave the laser field and continue to propagate in the plasma, and their energy is finally lost through collisions and other means. The vacuum heating mechanism often occurs in the process of laser-solid target interaction.

According to the capacitor model [11], considering the real driving electric field in the case of non-total reflection:

$$E_{\mathrm{d}} = [1 + (1 - \eta_{\mathrm{a}})^{1/2}] \cdot E_0 \sin\theta,$$

and assuming that all the electrons pulled out from the target surface return to the target and ignoring the effect of the $\boldsymbol{v} \times \boldsymbol{B}$ term, the laser conversion efficiency is:

$$\eta_{\mathrm{B}} = (1/\pi a_0) f[(1 + f^2 v_{\mathrm{os}}^2 \sin^2\theta/c^2)^{1/2} - 1](\sin\theta/\cos\theta), \qquad (3.3.3)$$

where $f = 1+(1-\eta_{\mathrm{a}})^{1/2}$ is the correction factor for the driving electric field. It can be seen that a larger incident angle and a stronger laser will enhance the vacuum heating mechanism and lead to more energy absorption.

3.3.4 $\boldsymbol{J} \times \boldsymbol{B}$ *Heating*

$\boldsymbol{J} \times \boldsymbol{B}$ heating [12] is similar to the vacuum heating mechanism in physical picture, both occurring in the process of laser interaction with plasma with a steep density boundary, and electrons are accelerated in the laser field to gain energy. Their main difference is that the main driving force of vacuum heating comes from the electric field perpendicular to the target surface, while $\boldsymbol{J} \times \boldsymbol{B}$ heating comes from the high-frequency term $\boldsymbol{v} \times \boldsymbol{B}$ (Lorentz force term) of the laser field radiation pressure, causing a nonlinear absorption. Similar to the discussion in Section 3.1.2, for a linearly polarized laser with the electric field form $\boldsymbol{E} = E_0(x)\sin\omega_0 t\widehat{\boldsymbol{y}}$, it will produce a longitudinal force:

$$\boldsymbol{f}_{\mathrm{p}} = -(m_{\mathrm{e}}/4)(\partial v_{\mathrm{os}}^2(x)/\partial x)(1-\cos 2\omega_0 t)\widehat{\boldsymbol{x}}. \tag{3.3.4}$$

The first term on the right-hand side of the above equation is the non-oscillating term, which is usually considered as the radiation pressure, providing a constant pressure to push electrons into the target and accelerate them, thereby causing the electron density distribution to be concave inward; the second term is the high-frequency oscillating term, and electrons oscillate at the interface at a frequency of $2\omega_0$ under the action of this force, thus being heated. It should be noted that the $\boldsymbol{J} \times \boldsymbol{B}$ heating mechanism exists in any non-circularly polarized laser, and it is most effective when the laser is normally incident on the target. When the electron oscillation velocity is relativistic, $\boldsymbol{J} \times \boldsymbol{B}$ heating becomes very important. For a normally incident circularly polarized laser on the target, the oscillating term in Eq. (3.3.4) is zero, and there is almost no $\boldsymbol{J} \times \boldsymbol{B}$ heating. Therefore, in radiation pressure acceleration, circularly polarized lasers are more inclined to be used to suppress this type of electron heating.

Generally, both vacuum heating and $\boldsymbol{J} \times \boldsymbol{B}$ heating will exist when a laser interacts with a plasma with a steep density boundary. When $2E_0 \sin\theta > v_{\mathrm{osc}} B_0$, vacuum heating dominates [10]; for a laser normally incident on the target, $\boldsymbol{J} \times \boldsymbol{B}$ heating dominates.

3.3.5 *Direct Laser Acceleration (DLA)*

Direct Laser Acceleration [13–17] (DLA) is the main electron acceleration mechanism in the interaction of long-pulse ($L > \lambda_{\mathrm{p}}$) intense lasers with near-critical density plasma ($0.1n_{\mathrm{c}} < n_{\mathrm{e}} < \gamma n_{\mathrm{c}}$). Here, L, λ_{p}, n_{e}, n_{c}, and λ are the spatial pulse width of the laser, the wavelength of the electron plasma wave, the plasma density, the critical density of plasma, and the relativistic factor of electrons, respectively. It is also listed here as a laser-plasma heating mechanism for a brief introduction.

In the DLA acceleration process, the relativistic laser usually meets the self-focusing condition [4], so the laser will first undergo a self-focusing process. At the same time, the radiation pressure of the laser pushes electrons to the surrounding transverse direction, forming a plasma channel in the subsequent propagation process (Figure 3.6(b)), and a strong self-generated quasi-static electromagnetic field is generated inside the channel (Figures 3.6(d)(e)). Electrons in the channel are transversely confined by the self-generated axial magnetic field and undergo transverse oscillation and longitudinal acceleration in the laser field (Figures 3.6(b)(c)). Since its energy mainly comes from the direct work of the laser transverse electric field, this acceleration process is called direct laser acceleration. When the transverse oscillation frequency of DLA electrons matches the laser field frequency in a special way [16,17], a large number of electrons will undergo resonant acceleration, and the energy and charge of the resonant electrons will be greatly enhanced. The specific process is not elaborated here. The DLA process contains rich physical processes, and the motion behavior and radiation characteristics of its electrons are still to be explored.

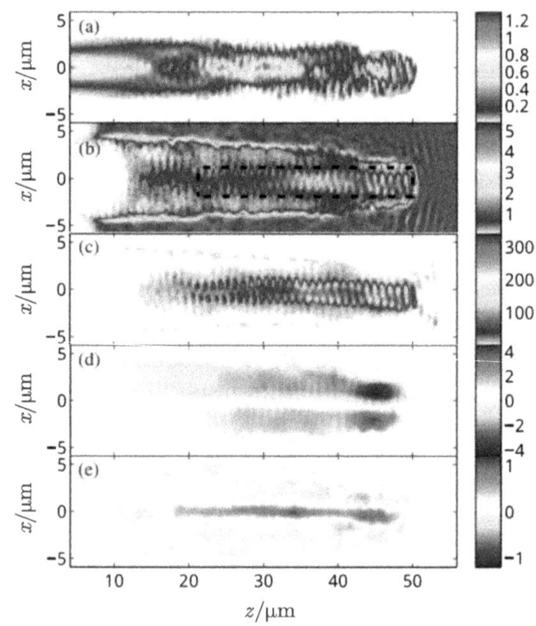

Fig. 3.6 Direct Laser Acceleration (DLA) in a critical density target [16]. (a) Instantaneous laser intensity, (b) electron density, (c) electron energy density, (d) self-generated azimuthal magnetic field, (e) self-generated axial magnetic field.

References

[1] Quinn M. N., McKenna P. Investigations of Fast Electron Transport in Intense Laser-Solid Interactions [D]. University of Strathclyde, 2011.

[2] Gibbon P. Short Pulse Laser Interactions with Matter: An Introduction [M]. London: Imperial College Press, 2005.

[3] Umstadter D. Relativistic Laser-Plasma Interactions [J]. *Journal of Physics D: Applied Physics*, 2003, 36: R151.

[4] Mora P., Antonsen J., Thomas M. Kinetic Modeling of Intense, Short Laser Pulses Propagating in Tenuous Plasmas [J]. *Physics of Plasmas*, 1997, 4: 217–229.

[5] Sun G. Z., Ott E., Lee Y., *et al.* Self-Focusing of Short Intense Pulses in Plasmas [J]. *The Physics of Fluids*, 1987, 30: 526–532.

[6] Wang H., Lin C., Sheng Z., *et al.* Laser Shaping of a Relativistic Intense, Short Gaussian Pulse by a Plasma Lens [J]. *Physical Review Letters*, 2011, 107: 265002.

[7] Bin J., Ma W., Wang H., *et al.* Ion Acceleration Using Relativistic Pulse Shaping in Near-Critical-Density Plasmas [J]. *Physical Review Letters*, 2015, 115: 064801.

[8] Kruer W. L. The Physics of Laser Plasma Interactions [M]. Boca Raton: CRC Press, 2019.

[9] Sentoku Y., Kruer W., Matsuoka M., *et al.* Laser Hole Boring and Hot Electron Generation in the Fast Ignition Scheme [J]. *Fusion Science and Technology*, 2006, 49: 278–296.

[10] Forslund D., Kindel J., Lee K. Theory of Hot-Electron Spectra at High Laser Intensity [J]. *Physical Review Letters*, 1977, 39: 284.

[11] Brunel F. Not-So-Resonant, Resonant Absorption [J]. *Physical Review Letters*, 1987, 59: 52.

[12] Kruer W. L., Estabrook K. $J \times B$ Heating by Very Intense Laser Light [J]. *The Physics of Fluids*, 1985, 28: 430–432.

[13] Pukhov A., Sheng Z.-M., Meyer-ter-Vehn J. Particle Acceleration in Relativistic Laser Channels [J]. *Physics of Plasmas*, 1999, 6: 2847–2854.

[14] Qiao B., Zhu S.-P., Zheng C., *et al.* Quasistatic Magnetic and Electric Fields Generated in Intense Laser Plasma Interaction [J]. *Physics of Plasmas*, 2005, 12: 053104.

[15] Arefiev A. V., Breizman B. N., Schollmeier M., *et al.* Parametric Amplification of Laser-Driven Electron Acceleration in Underdense Plasma [J]. *Physical Review Letters*, 2012, 108: 145004.

[16] Liu B., Wang H., Liu J., *et al.* Generating Overcritical Dense Relativistic Electron Beams via Self-Matching Resonance Acceleration [J]. *Physical Review Letters*, 2013, 110: 045002.

[17] Hu R., Liu B., Lu H., *et al.* Dense Helical Electron Bunch Generation in Near-Critical Density Plasmas with Ultrarelativistic Laser Intensities [J]. *Scientific Reports*, 2015, 5: 15499.

Chapter 4

Physical Mechanisms of Laser-Driven Ion Acceleration

At current laser intensities, lasers are not capable of directly accelerating ions. Instead, they must first transfer energy to electrons, which then generate an electrostatic field through charge separation to accelerate ions. The previous chapter introduced the physical mechanisms of laser heating of electrons. This chapter will delve into the mechanisms of ion acceleration when lasers interact with plasmas under various parameter conditions. The mechanisms and outcomes of ion acceleration in laser-ion acceleration experiments vary with experimental conditions, such as laser parameters (intensity, polarization, pulse width, focal spot size, laser contrast, etc.), target parameters (material, density, thickness, shape, etc.), and control conditions (incidence angle, focal distance, etc.). To date, researchers have proposed several ion acceleration mechanisms, including Target Normal Sheath Acceleration (TNSA), Radiation Pressure Acceleration (RPA), Collisionless Shock Acceleration (CSA), Breakout Afterburner Acceleration (BOA), and Coulomb Explosion (CE) acceleration, among others. Most of these mechanisms have been experimentally realized and verified. In recent years, several new acceleration mechanisms have also been proposed. This chapter primarily introduces the physical images and theoretical models of the common ion acceleration mechanisms. These theoretical models help further understand and analyze the interaction processes in laser-ion acceleration experiments. The scaling laws for ion energy provided by these models can also reasonably predict ion energy and beam quality characteristics. Additionally, this chapter briefly introduces the experimental progress of each acceleration mechanism to help readers gain a deeper understanding of the current state of laser-ion acceleration.

4.1 Target Normal Sheath Acceleration (TNSA)

4.1.1 *Physical Image of TNSA*

Target Normal Sheath Acceleration (TNSA) is a relatively mature acceleration mechanism that has been experimentally verified by many studies.

The basic physical image is as follows [1–3]: When a laser irradiates the front surface of a solid target, it generates relativistic superthermal electrons. As these electron beams propagate to the rear surface of the target, they establish an intense sheath electric field in the normal direction to the target back surface. This sheath electrostatic field can directly ionize atoms on the target back surface (protons mainly originate from contaminants such as water vapor and hydrocarbons on the target surface), and the resulting ion beam is accelerated in the electrostatic field and emitted at a certain solid angle, as shown in Figure 4.1.

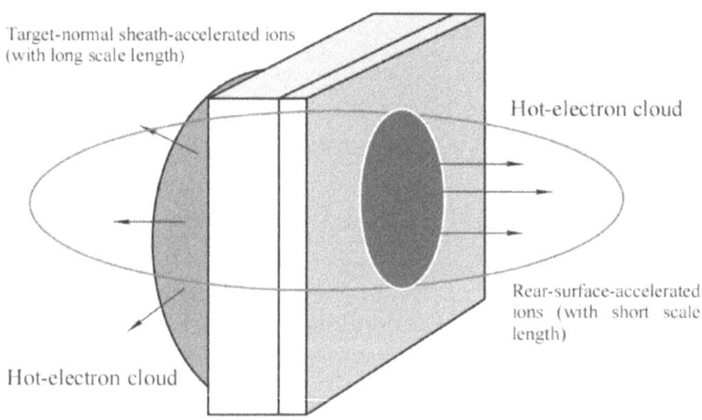

Fig. 4.1 Schematic diagram of TNSA mechanism [1].

4.1.2 *Scaling Laws of TNSA*

To explain the physical process of the TNSA mechanism, various theoretical models have been proposed, such as the isothermal free expansion model by Mora *et al.* [4, 5], the light ion acceleration model by Passoni *et al.*, and the two-particle component model by Tikhonchuk *et al.* Here, we mainly introduce the widely used isothermal free expansion model.

As shown in Figure 4.2, it is assumed that at the initial time $t = 0$, ions are stationary with a steep density profile, and the electron density

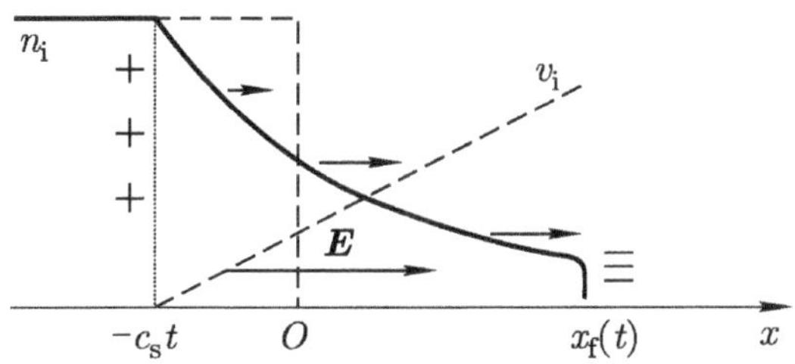

Fig. 4.2 Schematic diagram of vacuum free expansion model.

distribution follows a Boltzmann distribution, i.e.,

$$n_e = n_{e0} e^{e\Phi/T_e}, \quad x \in (-\infty, +\infty), \tag{4.1.1}$$

$$n_i = \begin{cases} n_{i0}, & x \in (-\infty, 0), \\ 0, & x \in (0, +\infty), \end{cases} \tag{4.1.2}$$

where Φ is the electrostatic potential, $\Phi(-\infty) = 0$, that is, $n_{e0} = Zn_{i0}$. The electrostatic potential satisfies Poisson's equation:

$$d^2\Phi/dx^2 = 4\pi(n_e - Z_i n_i). \tag{4.1.3}$$

In the formulas above, n_e and T_e represent the density and temperature of the superthermal electrons behind the target, respectively, while n_i and Z_i are the density and charge number of ions. When $t > 0$, ions begin to expand towards the rear of the target, while electrons still follow the Boltzmann distribution, and the Poisson equation remains valid. The characteristic length of the sheath electric field behind the target is the Debye length of the electrons:

$$\lambda_D = \sqrt{T_e/4\pi n_e e^2}. \tag{4.1.4}$$

Using the initial charge distribution and Poisson's formula, the strongest electric field at the initial moment is located at $x = 0$, with the electric field strength given by

$$E_{\text{sheath},0} = \sqrt{2}T_e/\sqrt{e_N}e\lambda_D = \sqrt{8\pi n_e T_e/e_N} = \sqrt{2/e_N}E_0, \tag{4.1.5}$$

where $e_N \approx 2.718\ 28$ is the Euler constant, and $E_0 = \sqrt{4\pi n_e T_e}$.

At any given moment, the electric field strength at the front end of the ion beam is

$$E_f = 2E_0/(2e_N + \omega_{pi}^2 t^2)^{1/2},$$

where $\omega_{pi} = \sqrt{4\pi n_{e0} e^2/m_i}$ is the ion plasma frequency.

Based on the ion continuity equation and the motion equation

$$\partial n_{\mathrm{i}}/\partial t + \nabla \cdot (n_{\mathrm{i}} v_{\mathrm{i}}) = 0, \tag{4.1.6}$$

$$ZeE = m_{\mathrm{i}} \mathrm{d}v_{\mathrm{i}}/\mathrm{d}t, \tag{4.1.7}$$

using the self-similar solution method [4], the electron density distribution can be obtained as

$$n_{\mathrm{e}} = n_{\mathrm{e}0} \exp(-x/c_{\mathrm{s}}t - 1), \tag{4.1.8}$$

the velocity of the ion front is

$$v_{\mathrm{f}} = 2c_{\mathrm{s}} \ln(\tau + \sqrt{1+\tau^2}), \tag{4.1.9}$$

and the position of the ion front is

$$x_{\mathrm{front}} = 2\sqrt{(2\mathrm{e_N})}\lambda_{\mathrm{D}}[\tau \ln(\tau + \sqrt{1+\tau^2}) - \sqrt{1+\tau^2} + 1],$$

where $\tau = \omega_{\mathrm{pi}} t/\sqrt{2\mathrm{e_N}}$ and $c_{\mathrm{s}} = \sqrt{Z_{\mathrm{i}} k_{\mathrm{B}} T_{\mathrm{e}}/m_{\mathrm{i}}}$, is the ion sound speed. When $\tau \gg 1$, $v_{\mathrm{f}} = 2c_{\mathrm{s}} \ln(2\tau)$, $x_{\mathrm{front}} = 2\sqrt{(2\mathrm{e_N})}\lambda_{\mathrm{D}}[\tau \ln(2\tau) - 1]$. Based on the ion density and corresponding velocity at different positions, the maximum energy and energy spectrum distribution of the accelerated ions can be obtained (where $E_0 = Z_{\mathrm{i}} k_{\mathrm{B}} T_{\mathrm{e}}$):

$$\varepsilon_{\mathrm{i}} = 2Z_{\mathrm{i}} T_{\mathrm{e}} [\ln(\tau + \sqrt{1+\tau^2})]^2, \tag{4.1.10}$$

$$\mathrm{d}N/\mathrm{d}E = (n_{\mathrm{i}0} c_{\mathrm{s}}/\sqrt{2EE_0}) \exp(-\sqrt{E/E_0}). \tag{4.1.11}$$

In the above model, the solid target is assumed to be infinitely thick. For targets of finite thickness, the isothermal expansion model is no longer fully applicable. However, based on the relationship between the laser pulse width and the time electrons move within the target, the thickness range of targets that meet the applicability conditions of the isothermal expansion model can be calculated. For example, for laser pulses with a pulse width of about 50 fs in experiments, the target thickness that meets the model's applicability conditions should be at least several micrometers.

In addition to the isothermal free expansion model, there are other models that can explain the ion acceleration process. Most of them also give the result that the maximum energy of ions ε_{i} is proportional to the temperature of the hot electrons T_{e}, that is,

$$\varepsilon_{\mathrm{i}} \propto T_{\mathrm{e}}. \tag{4.1.12}$$

Considering that the temperature of the hot electrons is approximately equal to the ponderomotive potential of the electrons (the energy electrons gain in the ponderomotive field)

$$T_{\mathrm{e}} = \Phi_{\mathrm{pond}} = m_{\mathrm{e}} c^2 \left(\sqrt{1+a_0^2} - 1\right), \tag{4.1.13}$$

the scaling law of ion energy in the TNSA acceleration mechanism can be expressed as

$$\varepsilon_{\mathrm{i}} \propto T_{\mathrm{e}} \propto \sqrt{1+a_0^2}-1 \propto I_0^{1/2}, \tag{4.1.14}$$

that is, for the TNSA mechanism, the cutoff energy of ions is approximately proportional to the square root of the laser intensity.

Although the theory of the TNSA mechanism is relatively mature at present, the ion spectra obtained in experiments usually show an exponential decay characteristic, and the proton energy and monoenergeticity are still far from the requirements for practical applications. Current research on the TNSA mechanism mainly focuses on how to improve the laser absorption efficiency by changing the target configuration and layout, thereby increasing the ion energy.

4.1.3 *TNSA Experiments*

Experimental research on laser ion acceleration [10–12] began in 2000. In the initial exploratory stage, there was some controversy over the specific acceleration mechanism of the emitted ions. In 2000, Clark *et al.* [11,13] used a laser with an intensity of 5×10^{19} W/cm^2 to irradiate a 125 μm aluminum target, producing approximately 10^{12} protons with energies above 2 MeV. When measuring the proton emission angle using a radiation-sensitive film, they observed a ring-shaped ion distribution (as shown in Figure 4.3). They believed that the high-energy protons originated from the front surface and were deflected by the internal ring-shaped magnetic field as they passed through the hot plasma. In the same year, Snavely *et al.* [12] used a super-strong laser with an energy of 48 J, power of 1 PW, and intensity of 3×10^{20} W/cm^2 to irradiate a 100 μm plastic target. They observed a collimated high-energy proton beam emitted from the rear surface of the target, with 12% of the laser energy being converted into 2×10^{13} protons with energies above 10 MeV. The proton spectrum exhibited a typical Maxwellian

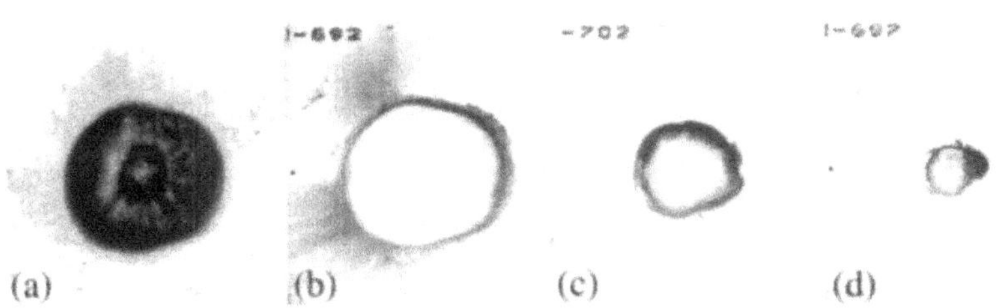

Fig. 4.3 Annular distribution of protons [11].

distribution with a cutoff energy as high as 58 MeV. When they further irradiated a 30° wedge target at a 45° angle, they found two separate proton beams emitted along the normal directions of the two right-angle edges of the wedge target.

In 2001, Wilks *et al.* [1] predicted that high-energy ions mainly originated from the acceleration by the sheath electric field behind the target; in the same year, Mackinnon's research on changing the plasma scale length on the target surface also concluded that protons originated from the rear surface [14]. More direct evidence came from the 2004 experiment by Allen *et al.* [15], who used an Ar ion gun to directly remove the contamination layer from the rear surface of the target material. They found that no high-energy protons were observed after removing the contamination layer, thereby determining that high-energy protons originated from the rear. Subsequently, the experiments by Fuchs *et al.* [16] also confirmed the source of high-energy ions, determining that ions from the rear had higher energy and lower divergence angles. Currently, most ion acceleration experiments worldwide are based on the TNSA mechanism.

In TNSA-dominated acceleration experiments, the laser intensity range is generally from 10^{18} W/cm^2 to 10^{20} W/cm^2, and the target materials include carbon-hydrogen targets, metal targets, polymer material targets, etc., with target thicknesses all in the micrometer range. Despite the fact that the above experiments based on the TNSA mechanism can produce protons with certain energies, the energy spectra of the accelerated ions basically show an exponential decay trend. The proton energy and monoenergeticity are still far from the beam quality required for practical applications. Moreover, the TNSA mechanism also has the disadvantage of low laser-to-high-energy ion energy conversion efficiency. Therefore, after the initial exploratory stage, in order to improve the acceleration quality, researchers have proposed a variety of optimized acceleration schemes.

Under unchanged laser conditions, the coupling efficiency between the laser and the target can be improved by modulating the target structure, thereby enhancing ion acceleration. In 2006, Schwoerer *et al.* from the Jena Quantum Optics Institute in Germany used a microstructured target with a miniature target placed behind a planar target (as shown in Figure 4.4) to obtain a quasi-monoenergetic proton beam with a central energy of 1.2 MeV and an energy spread of 25% under a laser intensity of 3×10^{19} W/cm^2 [17]. Similarly, Hegelich *et al.* also used a composite target to obtain quasi-monoenergetic C^{5+} ions with an energy spread of 17% and an average energy of 3 MeV [18]. The use of such planar miniature small targets

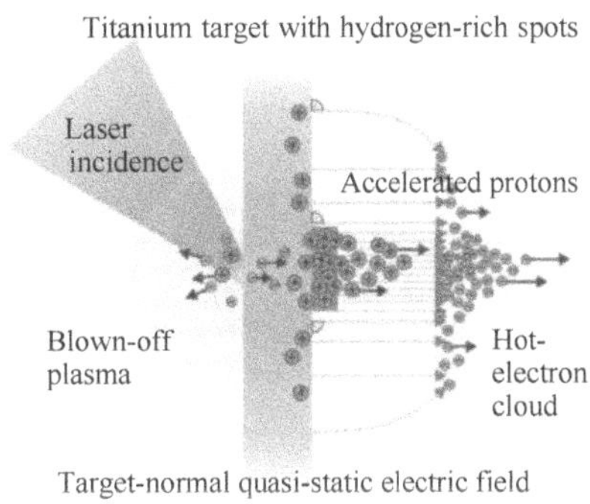

Fig. 4.4 Schematic diagram of obtaining quasi-monoenergetic protons by micro-structure target.

allows the locally accelerated ions to be in a more uniform sheath field, thereby reducing ion energy spread and improving beam quality.

Margarone *et al.* from the Czech Republic [19,20] spread polymer material nanospheres with diameters in the nanometer range on a planar target with a micrometer-scale thickness (as shown in Figure 4.5). The structural distribution of the miniature spherical target significantly increased the laser absorption efficiency. Under the same laser conditions, placing a nanosphere target in front of a planar target can almost double the proton energy. Passoni *et al.* from Italy placed an approximately 8 μm carbon foam target in front of a 0.75 μm planar aluminum target [21]. Under the same laser conditions, the carbon foam target can significantly improve the laser absorption efficiency and thereby increase the proton emission energy.

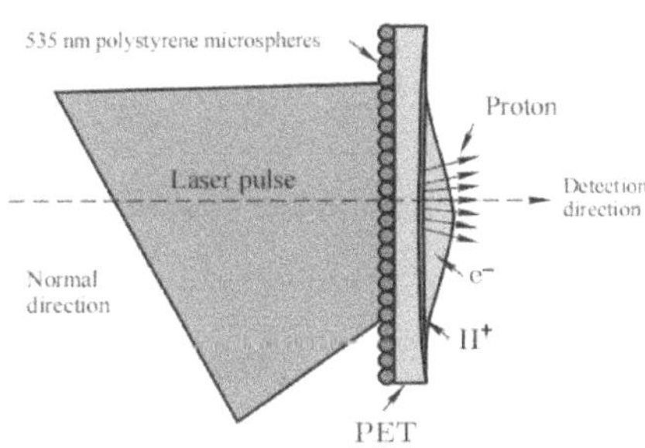

Fig. 4.5 Schematic diagram of nanosphere target acceleration [19].

In 2016, Wagner *et al.* from Germany [22] used a linearly polarized laser with an intensity of 2.6×10^{20} W/cm^2 to irradiate a 900 nm plastic target at an oblique angle, obtaining protons with the highest energy of 85 MeV. This experimental result represents the highest proton energy obtained in TNSA mechanism acceleration experiments to date.

4.2 Light Pressure Acceleration

4.2.1 *The Physical Picture of Radiation Pressure Acceleration (RPA)*

Thanks to the improvement of laser intensity and the maturity of target fabrication techniques, the radiation pressure acceleration model (RPA) [23–27] for laser-driven ultrathin targets has been proposed to achieve more efficient ion acceleration. Based on the physical process of laser interaction with ultrathin solid targets, the RPA acceleration process can be divided into two stages: the "hole-boring" stage and the "light-sail" stage. In the "hole-boring" stage, a high-intensity laser with a large focal spot irradiates the thin target. The laser's ponderomotive force compresses and expels electrons, creating a charge separation field. Ions are accelerated by this charge separation field, and the acceleration process then smoothly transitions from the "hole-boring" stage to the "light-sail" stage. In the "light-sail" stage, the expelled electrons and ions form a quasi-neutral plasma sheet that detaches from the target and acts as a "plasma mirror" driven by the laser. Since the plasma density is close to solid density, the laser is completely reflected, and the laser energy can be effectively converted into ion energy.

4.2.2 *Theoretical Models of Radiation Pressure Acceleration (RPA)*

When a laser irradiates an object moving at a velocity v perpendicularly, the radiation pressure exerted on the unit area is

$$P_{\mathrm{rad}} = R(2I_0/c)(1-\beta)/(1+\beta), \quad \beta = v/c, \tag{4.2.1}$$

where R is the reflectivity of the laser in the target's rest frame. If the thin target structure remains undamaged during the acceleration process, the motion equation of the thin target in one-dimensional ideal conditions can be written as [23, 28, 29]

$$\frac{\mathrm{d}p}{\mathrm{d}t} = R\frac{E_{\mathrm{L}}^2(t-X/c)}{2\pi n_{\mathrm{e}} l}\frac{\sqrt{p^2+m_{\mathrm{i}}^2c^2}-p}{\sqrt{p^2+m_{\mathrm{i}}^2c^2}+p}, \tag{4.2.2}$$

where m_i, n_i, n_e, and l are the ion mass, ion density, electron density, and target thickness, respectively. Correspondingly, the maximum energy of the accelerated ions can be given in the case of $R \approx 1$ [23]:

$$\varepsilon_i \approx m_i c^2 \left(\frac{3E_L^2 t}{8\pi n_e l m_i c}\right)^{1/3} = m_i c^2 \left(\mu s \frac{t}{T}\right)^{1/3}, \tag{4.2.3}$$

where $\mu = Zm_e/m_i$, $s = E_L^2/\sigma$, $\sigma = n_e l/n_c \lambda$ is the normalized surface density of the target, and T is the laser period. The above expression assumes an infinite acceleration time $t \to \infty$. If the acceleration time is finite, for example, for a laser pulse with a finite pulse width containing energy W_L, the maximum ion energy and the laser energy conversion efficiency at the end of the pulse can be expressed as [23, 29]

$$\varepsilon_{i,\max} = \frac{2W_L}{2W_L + N_i m_i c^2} \frac{W_L}{N_i}, \tag{4.2.4}$$

$$\eta = \frac{N_i \varepsilon_{i,\max}}{W_L} = \frac{2W_L}{2W_L + N_i m_i c^2}, \tag{4.2.5}$$

where N_i is the number of ions within the laser focal spot. From the above equations, it can be deduced that if the laser energy is much greater than the rest energy of the ions ($W_L \gg N_i m_i c^2$), most of the laser energy will be converted into ion kinetic energy $\varepsilon_{i,\max} \approx W_L N_i > m_i c^2$, $\eta \approx 1$. Thus, the scaling law of ion energy in the RPA mechanism can be estimated as: $\varepsilon_{i,\max} \propto I_0$, that is, the energy of the accelerated ions is proportional to the laser intensity to the first power. In contrast, the cutoff energy of ions in the TNSA mechanism is proportional to the square root of the laser intensity. Moreover, the energy conversion efficiency in the RPA mechanism can theoretically approach 100%, which is much higher than that in the TNSA mechanism. Therefore, it can be seen that the RPA mechanism is of great significance for accelerating high-energy ions using laser-plasma interactions.

4.2.3 *Conditions for Achieving RPA*

Although the RPA mechanism in laser acceleration has an extremely high laser-to-ion energy conversion rate, it also poses strict requirements for the acceleration conditions. To achieve stable RPA ion acceleration, the following conditions must be met:

(1) The laser intensity must be sufficiently high to expel electrons from the solid target as a whole. For targets with micrometer-scale thickness, this requires laser intensities far beyond what current laser technology can

achieve. For example, piston acceleration is an extreme case of the RPA mechanism, which not only requires extremely high focusing intensity but also demands that the usual Gaussian pulse in the time domain be transformed into a square wave pulse. These stringent conditions are temporarily difficult to realize [23].

(2) Circularly polarized lasers are preferable. As discussed in Section 3.3.4 regarding the ponderomotive force, when a linearly polarized laser pulse is incident perpendicularly on the front surface of a solid target, the oscillatory term in the ponderomotive force $f_{\mathrm{p}} \sim -\nabla E_0^2(x)[1-\cos(2\omega t)]$ causes electron oscillations, leading to intense electron heating and destruction of the acceleration structure. In contrast, using circularly polarized lasers, the ponderomotive force $f_{\mathrm{p}} \sim -\nabla E_0^2(x)$ does not contain the $\cos(2\omega t)$ oscillatory term, and electron heating is significantly suppressed. This is more conducive to the direct transfer of laser energy to the "plasma mirror," thereby achieving a higher energy conversion rate.

(3) The laser focal spot must be sufficiently large to maintain quasi-one-dimensional conditions throughout the entire acceleration process. The theoretical models mentioned above are all based on quasi-one-dimensional assumptions. If the laser focal spot is too small, more two-dimensional and three-dimensional effects need to be considered in the acceleration region, which will significantly reduce the laser conversion efficiency. Therefore, using ultra-Gaussian or flat-top pulses in the transverse direction of the incident laser is preferable to Gaussian pulses [26, 30, 31].

(4) Transverse instabilities [32] need to be controlled. In the RPA acceleration process, due to the use of high-intensity lasers and ultrathin targets, the development of transverse instabilities can easily destroy the acceleration structure and terminate the acceleration process. The next chapter will focus on the impact of transverse instabilities on the laser acceleration process.

4.2.4 *Phase-Stable Radiation Pressure Acceleration Region*

After the RPA method was proposed, it quickly attracted attention due to its high acceleration efficiency. However, theoretically, radiation pressure acceleration requires a laser power density of 10^{24} W/cm^2 [23], and the pulse waveform in the time domain must be a square wave, which is currently difficult to achieve experimentally. In order to realize this ideal and efficient ion acceleration mechanism in experiments and further improve the beam quality of ion beams, Yan Xueqing *et al.* discovered through theoretical analysis and numerical simulations that there exists a phase-stable region in the RPA mechanism: using moderately intense circularly polarized lasers interacting with nanometer-scale targets, the electron beam in the nanometer target is compressed towards the interior of the target as a whole. This results in a specific longitudinal electric field distribution in the solid target with a negative gradient characteristic (along the acceleration direction). This electrostatic field not only accelerates ions but also longitudinally bunches ions, theoretically effectively reducing the energy spread of the ion beam. The following introduces this phase-stable acceleration mechanism [27].

When a relativistic circularly polarized laser is incident perpendicularly on the surface of a solid target, the ponderomotive force, which lacks an oscillatory term, can compress electrons towards the interior of the target. A solid thin film target will gradually form two thin layers: one is an electron-depleted layer composed of positively charged ions, and the other is an electron compression layer with a negative charge. This forms a specific longitudinal electric field distribution as shown in Figure 4.6.

If the normalized laser electric field strength a_0, plasma electron density n, and target thickness D satisfy [27]

$$a_0(1+\eta)^{1/2} \sim (n_0/n_c)(D/\lambda_L), \tag{4.2.6}$$

the ponderomotive force will not be sufficient to expel electrons completely outside the target but will reach a balance with the electrostatic force. In Eq. (4.2.6), η is the laser reflection coefficient, n_0 is the initial plasma density, n_c is the critical plasma density, and λ_L is the wavelength of the laser in vacuum. When the laser is incident on the solid target, the electron

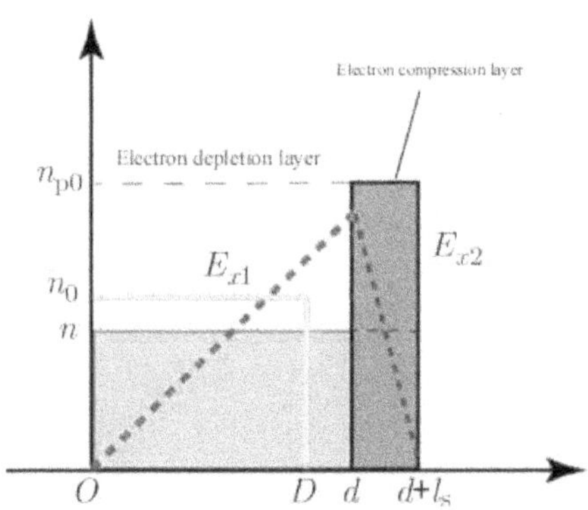

Fig. 4.6 Schematic diagram of electron density and ion density distribution [27].

compression layer will be slowly accelerated, continuously transferring momentum to the ions to be accelerated through the electrostatic force. As shown in Figure 4.6, for ions within the electron compression layer, the strength of the electrostatic force along the x-direction (E_{x2}) exhibits a negative gradient. This means that ions at the front experience a smaller acceleration field, while ions at the back experience a larger acceleration field. Therefore, ions that lag behind can catch up and overtake those in front, and this "overtaking" phenomenon can be observed in the ion phase space diagram shown in Figure 4.7. This is very similar to the phase-stable acceleration process in conventional radiofrequency accelerators [70], hence this acceleration mechanism is also known as phase-stable radiation pressure acceleration. Under this acceleration mechanism, the particle beam is compressed and bunched in the longitudinal phase space, as shown in the ion energy spectrum in Figure 4.8, with a significant reduction in beam energy spread. Meanwhile, as the laser continues to push the electron compression layer, the energy of the particles can also be continuously increased.

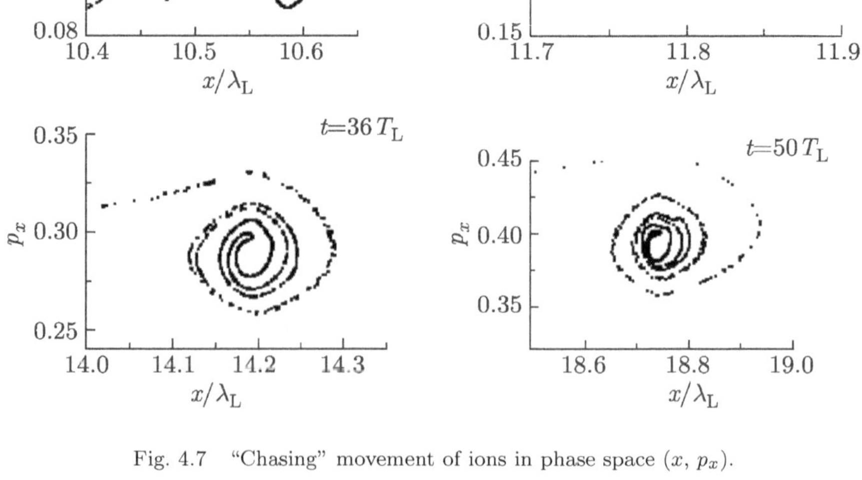

Fig. 4.7 "Chasing" movement of ions in phase space (x, p_x).

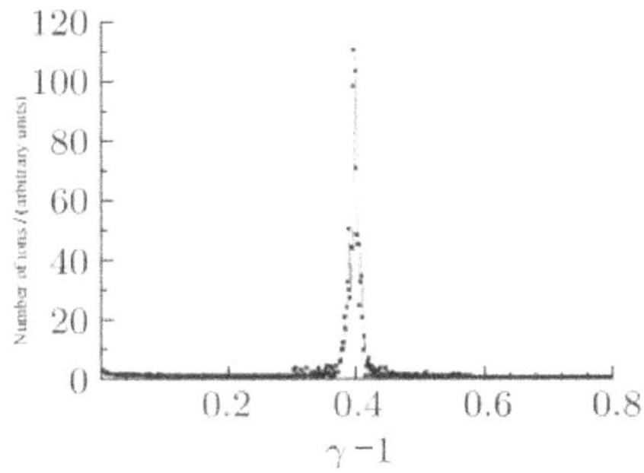

Fig. 4.8 Ion energy spectrum (γ is relativistic factor).

According to the above model, the number of accelerated ions can reach $a_0 n_0 \lambda_L (\pi \omega_L^2)$. If the laser beam radius is $w_L = 1\ \mu$m, the wavelength $\lambda_L = 1\ \mu$m, the liquid hydrogen target plasma density $n_0 \sim 10^{22}/\text{cm}^2$, and the normalized laser electric field strength $a_0 = 10$, theoretically, the number of high-energy protons produced can be in the order of 10^{11}. Phase-stable radiation pressure acceleration provides a possible solution for future compact table-top accelerators and is of great significance for the practical application of laser ion accelerators.

4.2.5 *RPA Experiments*

As mentioned above, in sheath field-dominated ion acceleration experiments, the target thickness is in the micrometer range, and the laser-to-ion energy conversion rate is usually low. Since Esirkepov *et al.* theoretically proposed the piston acceleration mechanism in 2004 [23], it has attracted widespread attention and research due to the good monoenergeticity of the resulting proton beam and the high laser energy conversion rate. However, due to the stringent requirements of the piston acceleration mechanism for laser parameters (extremely high intensity and short rise time) and the preparation of ultrathin targets, experimental conditions were not yet available. After 2008, people gradually discovered that as long as the laser radiation pressure and the electrostatic separation force of the solid thin film target are comparable, there exists a phase-stable acceleration region. Laser ion acceleration can be analogized to the principle of "sail" or "light-sail" acceleration, and the ion acceleration theory has made further breakthroughs [24,27,33–37]. These theories significantly reduced the stringent requirements of RPA for laser intensity and rise time, and since then, multiple teams worldwide have successfully conducted RPA experiments, with the main results introduced here.

In 2009, Henig, Steinke *et al.* from the Munich Quantum Optics Institute [38–40] first conducted experiments on radiation pressure-driven ion acceleration. They used a circularly polarized laser with an energy of 0.7 J and a focused intensity of 5×10^{19} W/cm^2, with an ultra-high contrast ratio of 10^{11}, to irradiate an ultrathin diamond-like carbon target perpendicularly. Under the optimal target thickness of 5.3 nm, they observed a monoenergetic peak structure of fully ionized carbon ions C^{6+} (as shown in Figure 4.9) [38]. At this time, electron heating was significantly suppressed, with the highest proton energy reaching 13 MeV and the highest carbon ion energy reaching 71 MeV. The laser energy conversion efficiency reached 10%, which is much higher than the results of previous thin-target ion acceleration experiments [39]. Subsequently, Kar *et al.* also reported experimental results on ultraintense laser-driven ion acceleration from multicomponent solid targets, using a 250 TW, 3×10^{20} W/cm^2 ultraintense laser to irradiate ultrathin targets, obtaining quasi-monoenergetic proton and carbon ion beams with peak energies of approximately 1–10 MeV/u (as shown in Figure 4.10) [41].

In 2013, Steinke *et al.* used a 5×10^{19} W/cm^2 circularly polarized laser to irradiate a 10 nm polymer material target, obtaining quasi-monoenergetic protons with a peak energy of $\sim$ 2 MeV. This part of the proton beam coupled about 6.5% of the laser energy, and they also clearly observed the multilayer interface structure of protons and carbon ions in the ion energy spectrum [41].

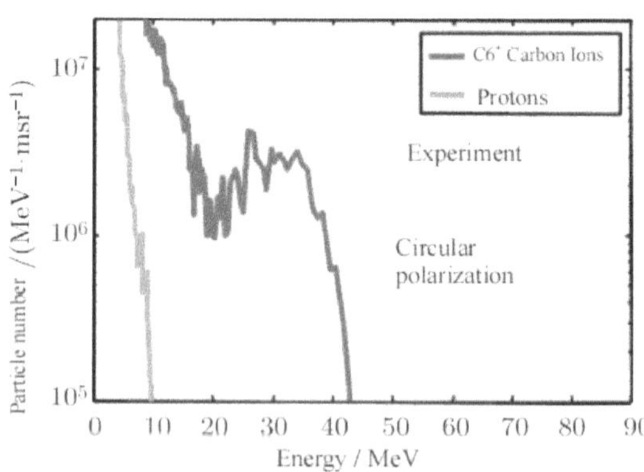

Fig. 4.9 Single energy peak structure of carbon ions observed in the experiment [38].

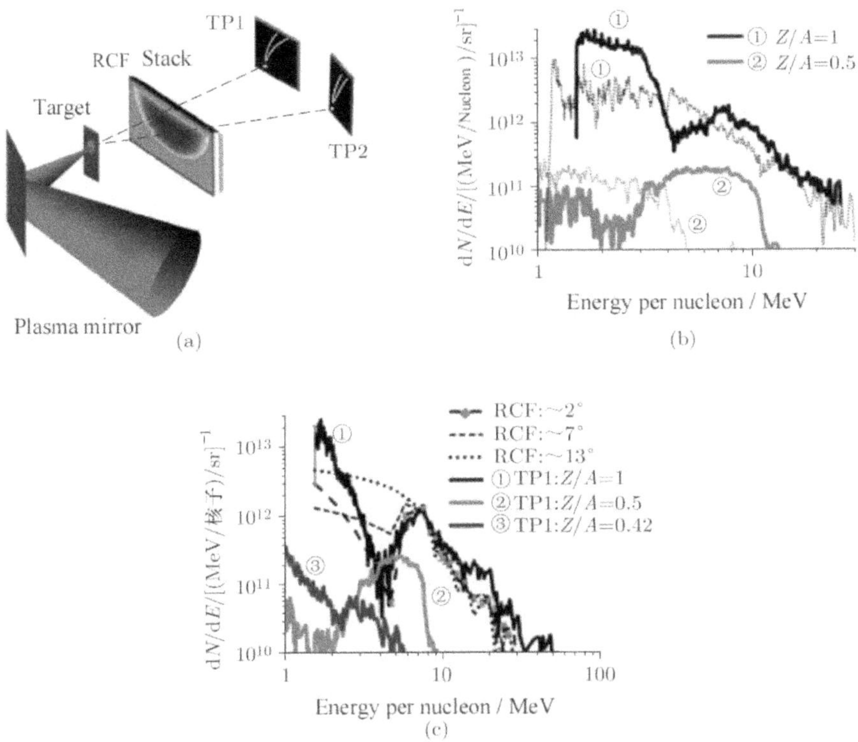

Fig. 4.10 Quasi-single energy peak structures of carbon ions and protons observed in the experiment. TP: Thomson spectrometer [41].

In 2015, Bin Jianhui from the University of Munich and Ma Wenjun from Peking University, among others, utilized the advantages of near-critical-density plasma lenses for laser shaping [42]. By placing a few micrometer-thick critical-density carbon nanotube structures in front of a nanometer-scale diamond-like carbon target [43], when a laser with an energy of 4 J and a focused intensity of 2×10^{20} W/cm^2 acted on the critical-density target, self-focusing and self-phase modulation non-linear effects shaped the laser pulse [42] — increasing its intensity and improving its contrast ratio, thereby enhancing ion acceleration. They obtained a quasi-monoenergetic carbon ion peak of $\sim$ 15 MeV/u (as shown in Figure 4.11). Subsequently, in 2019, the team of Ma Wenjun and Yan Xueqing from Peking University, in collaboration with Kim *et al.* from South Korea, once again leveraged the advantages of double-layer targets composed of critical-density plasma and diamond-like carbon targets,

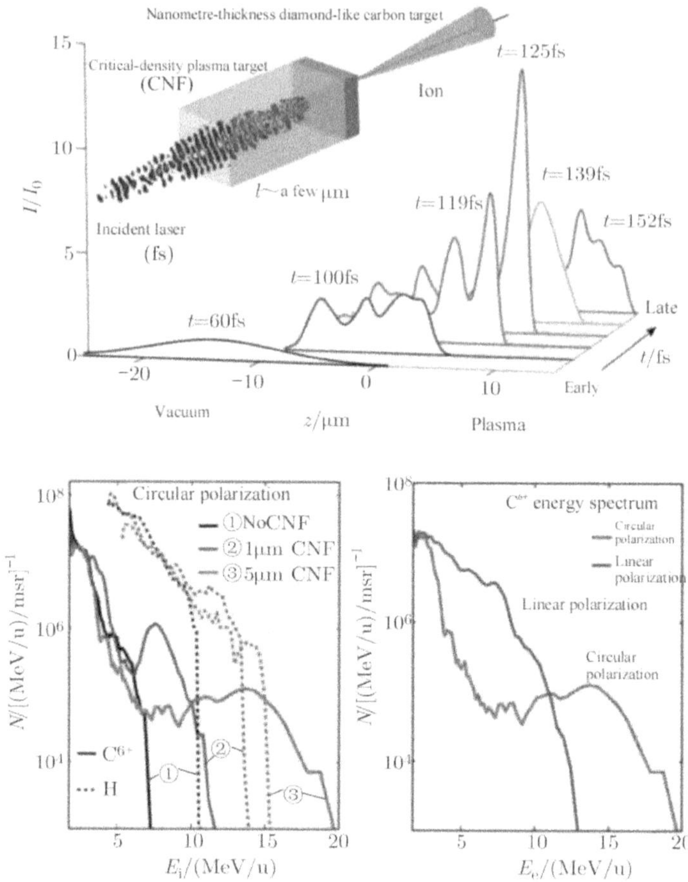

Fig. 4.11 Laser shaping and ion acceleration enhancement of carbon nanotube (CNF) critical density target [43].

achieving a new world record for carbon ion energy in femtosecond laser acceleration [44]. In the critical-density target, a large number of relativistic electrons were generated by direct laser acceleration. These electrons enhanced the target's rear sheath field in the subsequent shaped laser radiation pressure acceleration of carbon ions, realizing a cascade acceleration of heavy ions. The highest carbon ion energy reached 48 MeV/u [45]. The production of such high-energy ions confirmed the feasibility of cascade acceleration using double-layer targets and provided strong support for research in nuclear physics, high-energy density physics, and medical physics.

For proton acceleration, in 2016, Kim *et al.* from South Korea used a circularly polarized laser with a focused intensity of 6.1×10^{20} W/cm^2 to irradiate a polymer material nano-target. At the optimal target thickness of 15 nm, they obtained protons with a cutoff energy of 93 MeV [45]. However, the proton energy spectrum did not show a clear monoenergetic peak, possibly because the laser conditions did not meet the theoretical model requirements, and actual experiments were complicated by multidimensional effects.

In 2018, Higginson *et al.* [46] used a linearly polarized laser with an energy of 210 J and an intensity of 3×10^{20} W/cm^2 to irradiate an ultrathin plastic target at an oblique angle. Under the optimal target thickness of 75 nm, they produced protons with energies exceeding 94 MeV. Through PIC numerical simulations, they pointed out that the high-energy protons experienced a hybrid acceleration process of TNSA and RPA.

In summary, compared with target normal sheath acceleration (TNSA), radiation pressure acceleration (RPA) using high-contrast ultraintense lasers and nanometer-scale ultrathin targets has obvious advantages in improving ion energy, reducing ion energy spread, and increasing laser energy conversion efficiency. It is expected to become a mainstream direction for future ion acceleration experiments. However, experimental results also show that the actual ion or proton energies obtained through the RPA mechanism are still far below the theoretical and simulation predictions. Further in-depth research is needed to obtain higher-energy and better-quality ion beams through experiments.

4.3 Other Acceleration Mechanisms

Target normal sheath acceleration (TNSA) and radiation pressure acceleration (RPA) are the two most extensively studied ion acceleration mechanisms. In addition to these, there are several other ion acceleration mechanisms, which are briefly introduced here.

4.3.1 *Collisionless Shock Acceleration (CSA)*

Collisionless shock acceleration (CSA) was first proposed by Silva *et al.* in 2004 [47]. For a laser of a certain intensity, if the target is relatively thick, after the plasma is "bored" by the radiation pressure, the conditions for the "light-sail" stage cannot be met, and the collisionless electrostatic shock acceleration mechanism may occur [47–49].

In a quasi-static one-dimensional model, when the laser irradiates the target, the ponderomotive force compresses the electrons in the direction of laser propagation, forming an electron compression layer. The change in electron density distribution occurs while the ion density has not yet changed, resulting in a charge separation field. The ions left behind are accelerated forward by this charge separation field, forming an ion peak. This charge separation field has the distribution shown in Figure 4.12, and a potential barrier can be formed accordingly. Such a potential barrier propagating in the plasma can develop into an electrostatic shock (shock), with the shock velocity being a multiple of the ion acoustic speed, that is, $v_s = Mac_s$, where the Mach number Ma can be given by the Rankine–Hugoniot relation [50]:

$$n_i m_i v_s^2 = (1 + R) I_L / c, \tag{4.3.1}$$

$$Ma = \frac{v_s}{c_s} = \sqrt{\frac{\frac{N_2}{N_1}\frac{T_{e2}}{T_{e1}} - 1}{1 - \frac{N_1}{N_2}}}, \tag{4.3.2}$$

where N_1 and N_2 are the electron density distributions in the upstream (unshocked) and downstream (shocked) plasma regions of the shock, respectively, and T_{e1} and T_{e2} are the electron temperatures in the unshocked and shocked plasma regions, respectively.

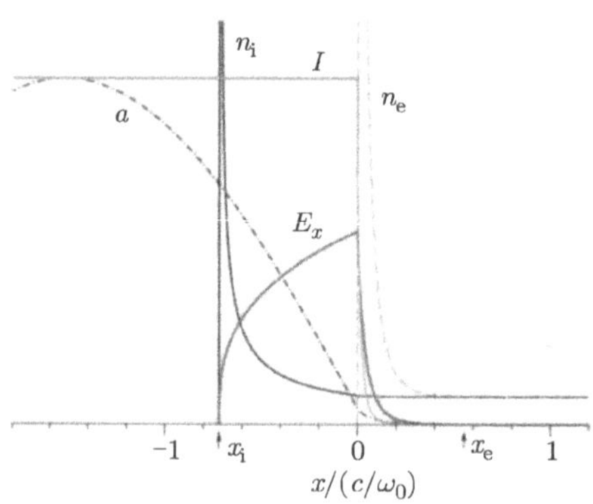

Fig. 4.12 Distribution of electron and ion density peaks before shock wave formation [50].

As shown in Figure 4.13, the plasma as a whole expands to the right at a velocity v_0, that is, the unaccelerated ions move to the right at an initial velocity v_0, while the shock moves to the right at a velocity $v_s = Mac_s + v_0$ (where v_s is v_{sh} and c_s is c_{s0} in the figure). In the reference frame of the

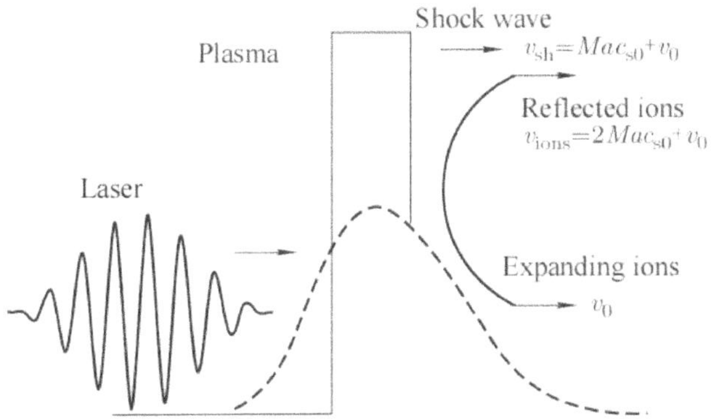

Fig. 4.13 Schematic diagram of shock acceleration mechanism [49].

moving shock, the stationary ions move to the left at a velocity $v_s - v_0$, and after encountering the shock, they are reflected and move to the right at a velocity $v_s - v_0$. Transforming back to the laboratory reference frame, the velocity of the ions reflected by the shock is v_i (denoted as v_{ions} in the figure) $= 2v_s - v_0 = 2Mac_s + v_0$. To form a shock, the Mach number $Ma = v_s/c_s$ cannot be too small. Forslund [51] gave the critical condition for the Mach number as $Ma \geq 1.6$. Since the ion velocity in shock acceleration is determined by the shock velocity, for a shock moving at a constant speed, a quasi-monoenergetic ion beam can be obtained. However, in actual situations, there are multiple factors that affect the quality of the accelerated ions: as the shock continuously accelerates ions, the energy of the shock also continuously decays; the shock velocity is related to the ion acoustic speed, which in turn is related to the electron temperature, and it is difficult to maintain a consistent shock velocity at different moments and in different regions; when extending the quasi-one-dimensional model to a multidimensional situation, multidimensional effects (such as lateral thermal diffusion in high-density accumulated regions, etc.) will also severely affect the shock velocity.

Shock acceleration can be realized in experiments using either solid targets or gas targets. In 2011, Haberberger *et al.* [52] used a 60 J carbon dioxide laser to irradiate a hydrogen gas target and observed the formation of a plasma shock in the experiment (as shown in Figure 4.14), obtaining a proton beam with a peak energy of 20 MeV and an energy spread of about 1% [52]. In 2015, H. Zhang *et al.* from the Shanghai Institute of Optics and Fine Mechanics used a laser with a focused intensity of 2×10^{19} W/cm^2

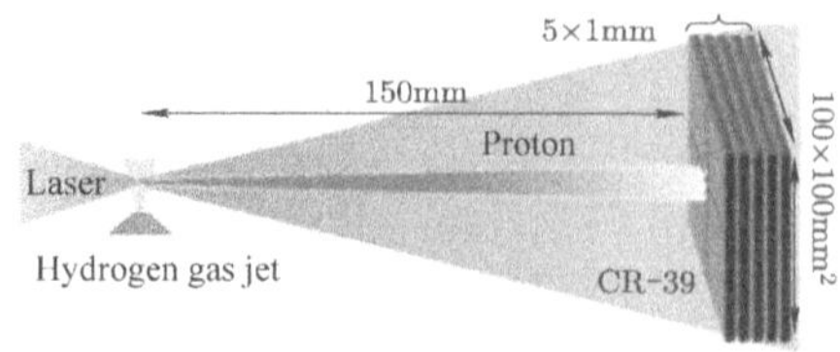

Fig. 4.14 Shock acceleration (CSA) experiment diagram.

to irradiate a diamond-like carbon film target. The pre-pulse heated the target front and formed a near-critical-density preplasma. When the main pulse acted on the preplasma, it created an electron density peak in the plasma, and the shock generated by this electron density peak accelerated the carbon ions in the target, resulting in a quasi-monoenergetic carbon ion beam with a peak energy of about 7.5 MeV [53].

4.3.2 *Breakout Afterburner Acceleration (BOA)*

Breakout afterburner acceleration (BOA) [54, 55] was first proposed by Yin *et al.* from the Los Alamos National Laboratory (LANL) in the United States. The BOA mechanism employs a linearly polarized laser with an intensity of about 10^{21} W/cm^2 and a target thickness in the nanometer range (10–500 nm). The acceleration process can be divided into three stages:

(1) The laser pulse is incident on a high-density solid target, and due to the oscillatory term in the ponderomotive force, the laser continuously heats electrons at the front surface of the target.

(2) Enhanced TNSA acceleration stage. Since the target is relatively thin, the cold electrons in the target that have not been sufficiently heated can reflux and be heated by the laser to produce superthermal electrons, further increasing the electron temperature. Since the sheath field strength is proportional to the square root of the electron temperature, a higher electron temperature can induce a stronger rear sheath field. At this time, although the plasma is in a high-density state, its electron skin depth can be comparable to the target thickness, and the laser begins to partially penetrate the target.

(3) BOA acceleration stage. As the electron density decreases with the expansion of superthermal electrons, when the electron density n_e drops to $n_e \approx \gamma n_c$, the plasma reaches a relativistically induced transparency state. At this point, the laser and electrons move forward together, generating a

large-amplitude local longitudinal electric field at the rear surface of the thin target. This electric field peak moves synchronously with the ions and continuously accelerates them, which is the main ion acceleration stage of BOA. Compared with the TNSA mechanism, the BOA mechanism can achieve higher ion energies and laser energy conversion efficiency.

Since the BOA mechanism uses a target with a thickness comparable to the plasma skin depth, the axial electron layer will be nearly depleted. A large number of electrons will accumulate in the direction perpendicular to the polarization axis, resulting in high-energy ions mainly distributed on both sides of the central axis (Figure 4.15). This is significantly different from the ion beam in the TNSA mechanism, which is mainly concentrated in the direction perpendicular to the target rear [55].

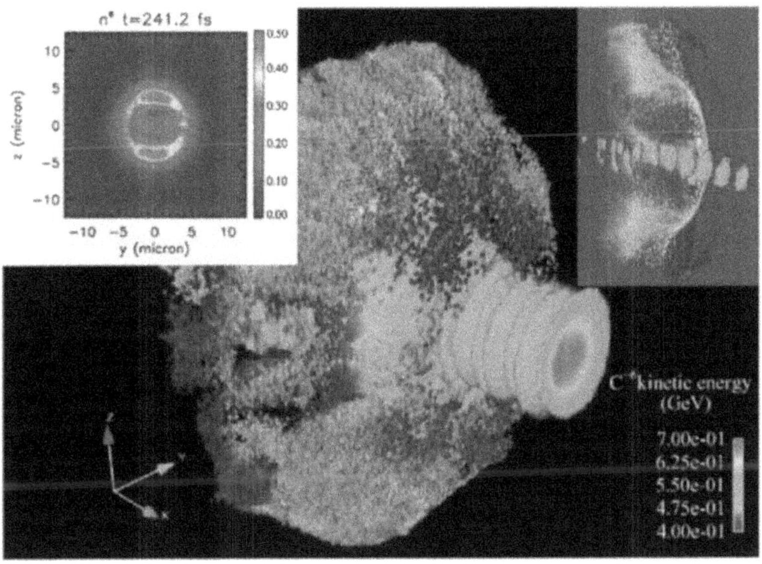

Fig. 4.15 BOA accelerated high-energy ions are mainly distributed in directions perpendicular to laser propagation and polarization [55].

Compared with the TNSA mechanism, the BOA mechanism requires a thinner target under the same laser conditions and more sufficient electron heating. In 2013, Hegelich *et al.* used a laser with an energy of 90 J and a focused intensity of 2×10^{20} W/cm^2 to irradiate a thin target. They found that ion acceleration was most effective when the target was relativistically transparent, and this state was maintained until the electron density in the target decreased below the critical density. Under the optimal target

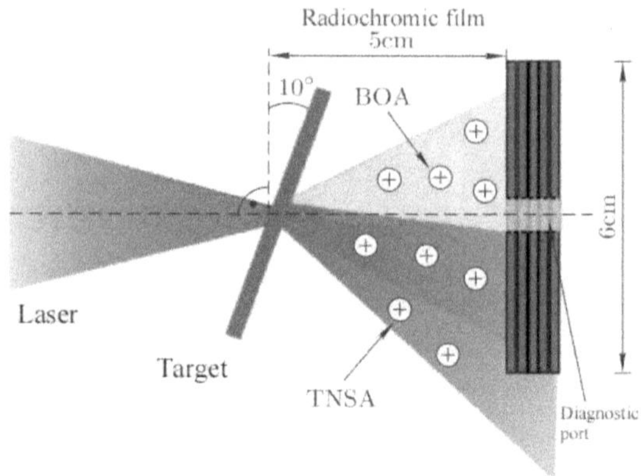

Fig. 4.16 Acceleration mechanism of TNSA and BOA [57].

thickness, they obtained protons with an energy of 37.8 MeV [56]. In 2015, Wagner *et al.* used a laser with a focused intensity of 6×10^{20} W/cm^2 to irradiate a polymer material target at a 10° angle. They observed two proton beams with different angular distributions. As shown in Figure 4.16, the proton beam emitted in the normal direction of the target rear had a cutoff energy of about 30 MeV, which was independent of the target thickness and was considered to be accelerated by the TNSA mechanism; while the proton beam observed in the lateral region of the central axis had a maximum energy that changed with the target thickness, and the cutoff energy reached 61 MeV at the optimal target thickness of 750 nm, which was considered to be accelerated by the BOA mechanism [57].

4.3.3 *Coulomb Explosion (CE) Acceleration*

The acceleration mechanisms discussed earlier mainly involve laser interactions with planar solid targets. In laser ion acceleration experiments, some researchers have also attempted to use other forms of targets, such as cluster targets (a material structure intermediate between atoms, molecules, and macroscopic solid substances). The size of clusters is in the range of a few hundred nanometers to micrometers, with a density comparable to that of solids. However, there are gaps between clusters, and when the laser interacts with cluster targets, the laser ionizes the clusters into plasma and expands them uniformly, resulting in a plasma density much lower than that of solids. Compared with planar solid targets, cluster targets have a higher laser energy absorption efficiency.

Coulomb explosion (CE) acceleration is mainly driven by electrostatic pressure, which is different from the thermal pressure drive in the target normal sheath acceleration (TNSA) mechanism and the radiation pressure drive in the radiation pressure acceleration (RPA) mechanism. Coulomb explosion acceleration [58–60] is the main mechanism for generating high-energy ions when a laser irradiates a cluster target. The basic physical process is as follows: the laser irradiates the cluster target, and most of the laser energy is first deposited into the electrons in the cluster. These hot electrons expand outward, forming a charge distribution with positive charge in the center and negative charge around, which excites a radial electric field and accelerates ions. It should be noted that the electrostatic field experienced by ions comes partly from the attraction of expanding electrons and partly from the repulsion of surrounding ions.

Considering the case where electrons are completely expelled from the target and do not return to it, the laser intensity required to strip all electrons from the cluster atoms is

$$a_0 \geq 2\pi n_{\mathrm{e}} R_0 / 3 n_{\mathrm{c}} \lambda_0. \tag{4.3.3}$$

The maximum energy of ions accelerated from the cluster surface can be estimated by

$$E_{\max} = 4\pi e^2 Z_{\mathrm{i}} R_0^3 n_{\mathrm{e}} / 3 = 3 Z_{\mathrm{i}} m_{\mathrm{e}} c^2 a_0^2 n_{\mathrm{c}} / n_{\mathrm{e}}. \tag{4.3.4}$$

Physically, ions located at the same radius experience forces of the same magnitude. If the cluster target contains multiple types of ions and a certain type of ion is distributed on the surface, quasi-monoenergetic particles can be obtained.

In Coulomb explosion acceleration experiments, the targets used are generally cluster targets or spherical microtargets. Ostermayr and Sokollik *et al.* from the University of Munich [61,62] proposed the scheme of isolated spherical microtargets. These spherical microtargets can be completely suspended and self-supported by an external magnetic field, thereby avoiding the current effects on ion acceleration that would be caused by other support methods. In their 2016 experiment (as shown in Figure 4.17), they used a laser with an intensity of $(2 \sim 3) \times 10^{20}$ W/cm^2 to irradiate isolated spherical microtargets with diameters ranging from 520 nm to 19.3 μm. When the target diameter was about 10 μm (comparable to the full width at half maximum of the laser), they obtained protons with a maximum energy of 25 MeV [61].

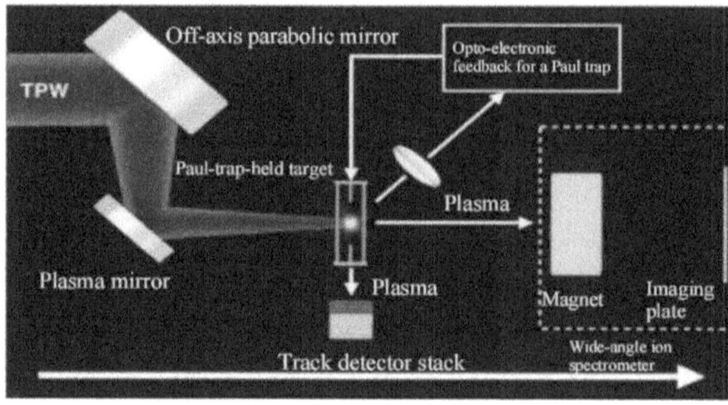

Fig. 4.17 Schematic diagram of Coulomb explosion acceleration experiment with isolated spherical small target [61].

4.4 Cascaded Acceleration and Post-Acceleration

4.4.1 *Cascaded Acceleration*

Laser accelerators have extremely high acceleration electric field gradients, and it has been experimentally verified that PW lasers can accelerate protons to nearly 100 MeV within a sub-millimeter scale. According to the proton energy scaling law, obtaining higher energies relies on higher-power lasers. Although laser technology is developing and experimental demands are increasing, higher-power ultraintense laser facilities are being constructed and will be put into use in the near future. However, this also significantly increases the cost of building laser ion accelerators.

Drawing on the ideas of conventional accelerators, a cascade acceleration scheme can be adopted. For example, combining traditional radiofrequency accelerators with laser accelerators to enhance proton energy. However, such a scheme loses the advantages of compact layout and cost-effectiveness of laser accelerators. Over the past decade, researchers have proposed various multi-stage coupled acceleration schemes combining different laser acceleration mechanisms. Compared with single-stage acceleration schemes, cascade acceleration relaxes the requirements for laser energy and intensity at each stage, avoiding damage and destruction to optical materials caused by high-intensity lasers. Moreover, different plasma and laser combinations can be selected at each stage of cascade acceleration, offering greater flexibility and controllability.

In 2010, Pftenhauer [63] proposed an early cascade acceleration scheme: a two-stage laser acceleration scheme. As shown in Figure 4.18, in the

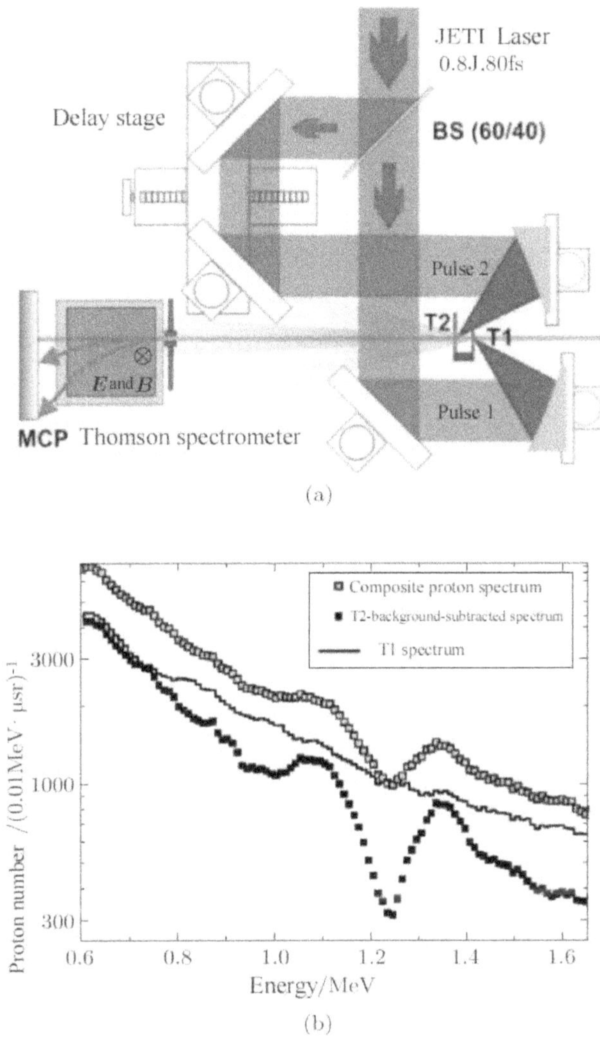

Fig. 4.18 Multi-stage TNSA acceleration scheme [63]. (a) Experimental diagram; (b) Proton spectral distribution.

multi-stage acceleration scheme, the first laser interacts with the first target (T1) to produce MeV proton beams using the TNSA acceleration method. When the protons propagate forward to the second target (T2), the second laser interacts with the second target to establish a sheath electric field at the rear of the target, modulating the feature spectrum of the primary proton beam. Only a portion of the protons in the second target is affected: protons behind the target are further accelerated, while those in front of the target are decelerated by the Coulomb potential barrier. By changing the delay time between the two lasers, the ion energy spectrum distribution can be manipulated. Despite this, in experiments, the close distance between the two solid targets and the difficulty in optimizing and controlling the timing of the two lasers result in a weak sheath electric field, and no significant proton energy enhancement was observed.

There are also some cascade acceleration schemes involving microstructures. In 2006, Toncian *et al.* [64] proposed a scheme using microtubes to focus and energy-select proton beams. A picosecond laser pulse acting on a hollow microtube generates a transient radial electric field, which can be used for focusing. As shown in Figure 4.19, the second laser (CPA2) strikes the wall of the metal tube, causing the inner wall to produce hot electrons that expand inward, thereby generating an inward space charge field that can focus the proton beam. In 2017, Sheng Zhengming *et al.* from Shanghai Jiao Tong University [65] discovered in simulations that when an ultraintense laser acts on the outer wall of a 1.6 μm hollow metal tube, in addition to the radial electric field, an axial electric field of the order of TV/m is also generated, which can further accelerate the protons. This

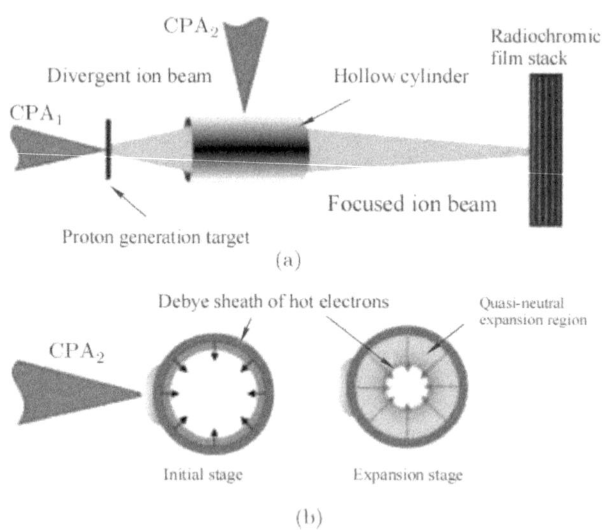

Fig. 4.19 Micron tube cascade acceleration. (a) Schematic diagram of micron tube; (b) Formation of charge separation field [68].

scheme ultimately increased the highest energy of the injected protons from 10 MeV to 17 MeV, and the energy spread from 10% to 4.9%. Increasing the length of the microtube may further improve the proton cutoff energy and collimation. Generally speaking, the greater the total laser energy, the higher the cost of the laser. Multi-stage cascade acceleration requires distributing the laser energy to multiple driving beams, which demands a larger total laser energy. If the ion energy gain cannot be increased, from a cost perspective, it may not be as advantageous as single-stage acceleration.

In 2019, Qiao Bin *et al.* proposed a scheme for all-optical ion cascade acceleration driven by ultraintense laser pulses [67], in which different acceleration stages are independent. As shown in Figure 4.20(a), in the first stage, a PW, picosecond laser[1] interacts with the first section of the microtube, forming a critical density plasma inside the tube, which can produce quasi-monoenergetic, highly collimated proton beams. In the second stage, two TW, femtosecond lasers act on the wings of the second tubular structure, driving surface currents along the tube wall to generate electromagnetic field pulses for proton cascade acceleration. The simulation results shown in Figure 4.20 indicate that the cutoff energy of the primary accelerated protons is 123 MeV, which increases to 151 MeV after the first cascade and to 181 MeV after the second cascade, with each stage providing an energy gain of about 30 MeV and an acceleration gradient of 0.3 TV/m. The total energy conversion efficiency is 2.1%. This acceleration mechanism can produce high-flux ion beams with very low energy spectrum divergence. Adopting kJ-level picosecond lasers to produce high-flux ion beams can partially compensate for the inevitable particle losses during cascade acceleration, which is conducive to the multiple cascade acceleration of proton beams. This scheme proposes a valuable cascade idea. Considering the high energy requirements for the laser, it is difficult to conduct experiments under current conditions.

4.4.2 *Post-Acceleration*

In the TNSA mechanism, since the ion velocity is low, the large amount of energy from the generated superthermal electrons is difficult to effectively transfer to the ions, and most of it escapes from the rear of the target, resulting in a laser-to-proton beam energy conversion efficiency typically less

[1]This book describes generally referred to as "picosecond" and "femtosecond" lasers, and in terms of power they are often referred to as "PW" and "TW" lasers, which is in line with the customary notation in this field of application.

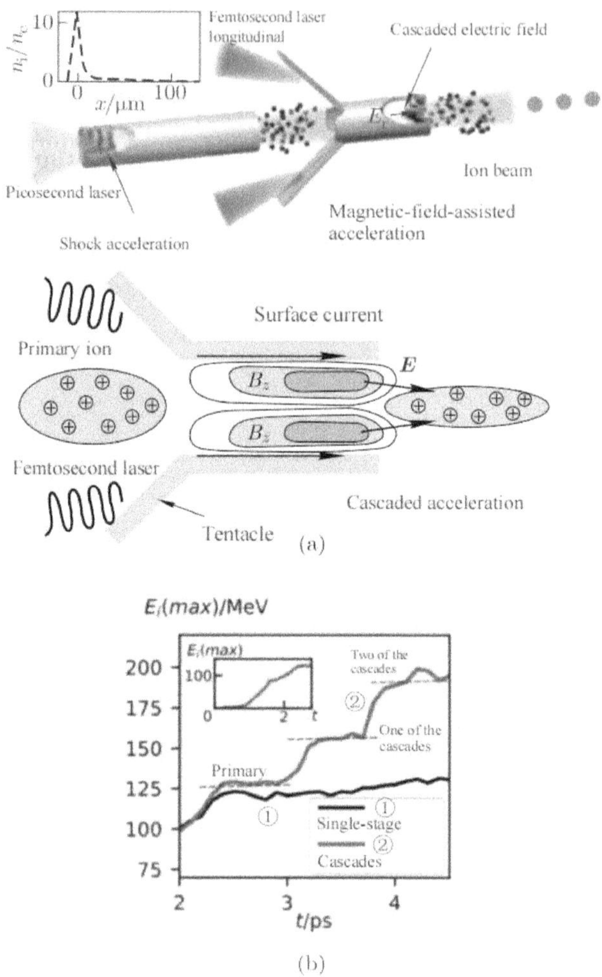

Fig. 4.20 All-optical ion acceleration cascade scheme. (a) Protons are accelerated by micron tubes; (b) Evolution of proton cutoff energy [66].

than 1% for femtosecond ultraintense lasers. To improve the energy conversion efficiency, the spiral slow-wave post-acceleration mechanism shown in Figure 4.21(a) has been proposed [67]. This mechanism uses a spiral slow-wave structure to collect and guide the electromagnetic pulse (EMP) generated by target irradiation, slowing down the electromagnetic pulse traveling at the speed of light to match the velocity of the proton beam. This principle is similar to the spiral waveguide accelerator developed by

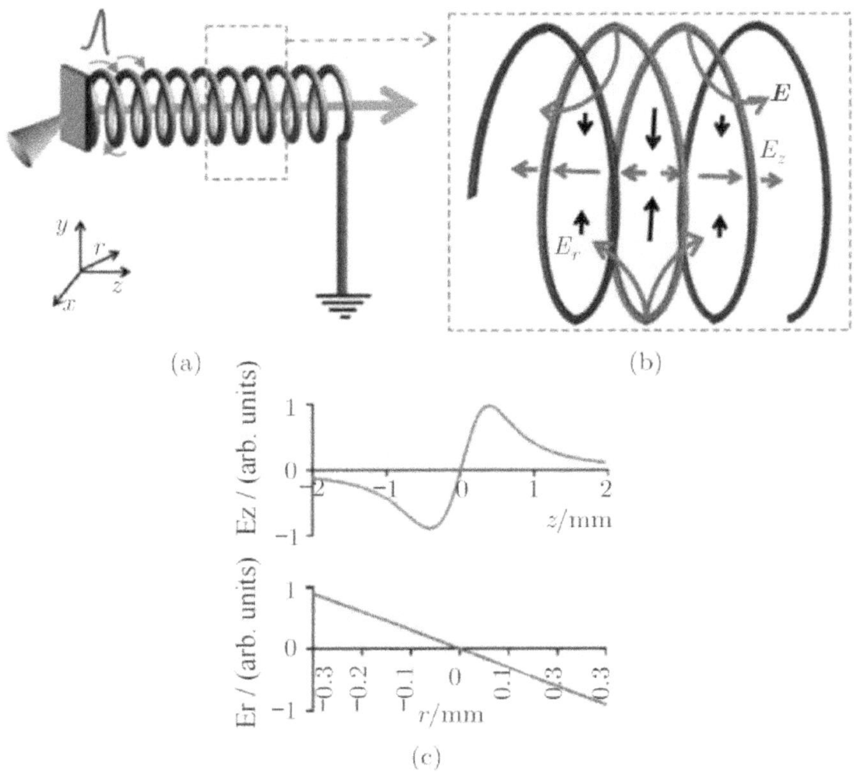

Fig. 4.21 Post-acceleration. (a) Experimental setup; (b) Spatial distribution of electric field; (c) Longitudinal and transverse electric field distribution [67].

Peking University in the 1980s [70]. The electric field distribution of the electromagnetic pulse is shown in Figures 4.21(b) and 4.21(c). If the protons are in the appropriate phase, they can obtain post-acceleration and focusing effects.

In 2016, Kar *et al.* used spiral slow-wave post-acceleration to achieve an energy gain of about 7 MeV [67]. The experimental RCF results are shown in Figure 4.22. Figure 4.22(a) shows that without the spiral slow-wave structure, under a laser pulse width of 30 fs and an intensity of 10^{20} W/cm^2, a metal target can produce a proton beam with a cutoff energy of 7.5 MeV. When using the spiral slow-wave structure, as shown in Figure 4.22(b), the cutoff energy is increased to 14 MeV, and the proton beam divergence is effectively reduced. The dashed circle in the figure indicates the size of the solenoid. In the experiment, a 200 TW laser was used, and the

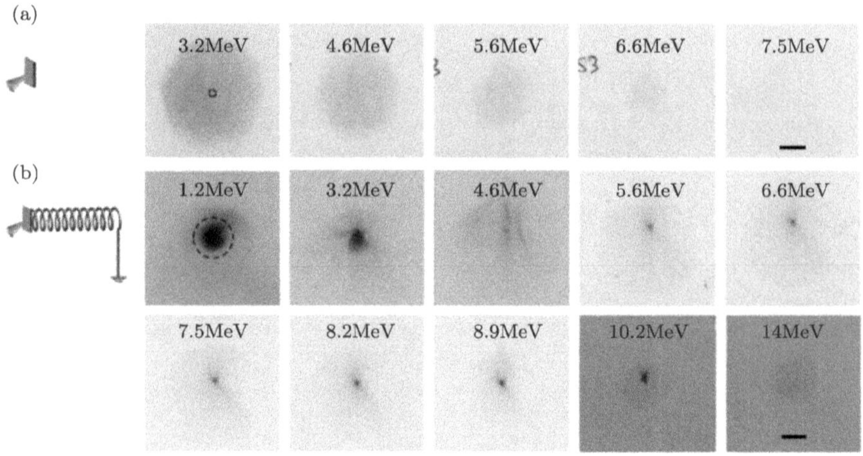

Fig. 4.22 Accelerated RCF data after spiral slow wave structure [67].

acceleration gradient in the post-acceleration section reached 0.5 GV/m. Theoretically, electromagnetic field pulses excited by multiple laser beams can also be used to further realize the manipulation and optimization of MeV ion beams through multiple cascade accelerations.

In 2020, Kar *et al.* used a spiral slow-wave structure on the VULCAN laser to increase the cutoff energy of a proton beam from 35 MeV to 48.9 MeV [68]. They also found that increasing the length of the solenoid appropriately can increase the energy gain, as shown in Figure 4.23.

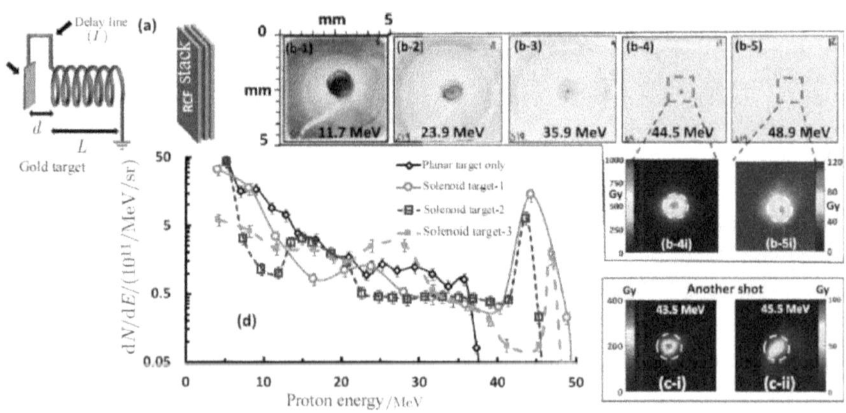

Fig. 4.23 Experimental results of solenoid acceleration after high power laser [68].

In the spiral slow-wave structure, the proton energy and velocity continue to increase, but the wave speed v_p of the electromagnetic pulse remains constant. This leads to a mismatch between the proton velocity and the wave speed, causing the energy gain to reach a bottleneck. The solution is to change the pitch to make the wave speed v_p consistent with the proton velocity, thereby increasing the effective distance of post-acceleration and further enhancing the gain, as shown by the solid line in Figure 4.24(a). In addition, they also simulated the effect of proton beams after passing through two solenoid structures. As shown in Figure 4.24(b), under a PW laser, the primary proton beam emitted by the thin-film target has a cutoff energy of 40 MeV. After passing through one solenoid structure, it reaches 70 MeV. After two stages of cascade acceleration, it reaches 100 MeV. It can be seen that the solenoid cascade acceleration scheme on the PW laser has the potential to produce high-energy proton beams with energies of 100–200 MeV.

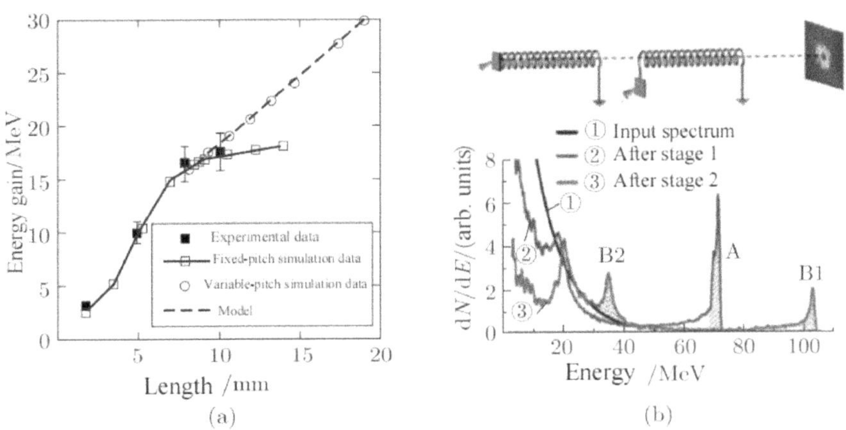

Fig. 4.24 (a) Variation of energy gain with solenoid length [68]; (b) experimental design and simulation results of solenoid cascade acceleration [67].

4.5 Discussion on Acceleration Mechanisms

This chapter has introduced the theoretical models and experimental research on laser-driven ion acceleration mechanisms. Overall, among all ion acceleration schemes, target normal sheath acceleration (TNSA) is the mechanism that is easiest to realize in laboratories and is most widely used.

However, this mechanism has disadvantages such as low ion energy, low laser-to-ion energy conversion efficiency, and large beam divergence angles, which limit its practical application range. Meanwhile, research on other acceleration mechanisms (such as breakout afterburner acceleration (BOA), collisionless shock acceleration (CSA), Coulomb explosion (CE) acceleration, etc.) has also been widely carried out. Relatively speaking, radiation pressure acceleration (RPA) is undoubtedly the most promising acceleration mechanism in terms of ion quality and laser conversion efficiency. However, this mechanism has stringent requirements for laser pulse conditions such as contrast ratio and temporal-spatial waveform, and therefore there are relatively few successful experimental results at present.

The interaction between laser and plasma is a complex process that depends on a large number of parameters. The physical processes corresponding to different laser normalized intensity a_0 and different target thickness conditions are shown in Figure 4.25. The boundaries between the various mechanisms in the figure are not strict. This is because, in real experiments, although different acceleration stages often correspond to different acceleration mechanisms, the transitions between the stages are

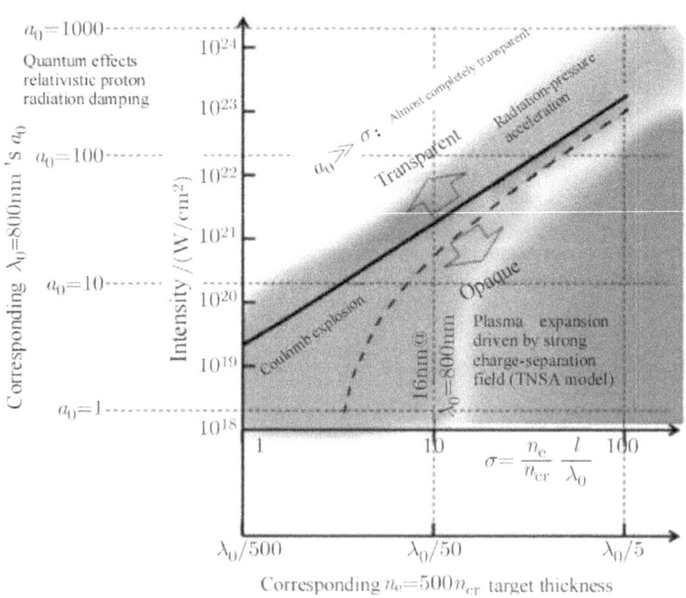

Fig. 4.25 Relationship between ion acceleration mechanism and experimental conditions, n_{cr} is the critical density [69].

continuous, and multiple acceleration mechanisms may also act simultaneously within one acceleration stage. Therefore, in analyzing simulation or experimental results, it is necessary to be very cautious in determining the acceleration mechanisms in the acceleration stages. In addition to referring to ion energy, other detection results should also be combined-such as electron energy spectra, laser absorption, reflection, or transmission rates, etc. For example, ions observed at the target rear may originate from either the front or rear of the target, and the high-energy part mainly comes from the rear. Therefore, when analyzing simulation or experimental results, it is necessary to be very cautious in determining the acceleration mechanisms in the acceleration stages. In addition to referring to ion energy, other detection results should also be combined-such as electron energy spectra, laser absorption, reflection, or transmission rates, etc.

References

[1] Wilks S. C., Langdon A. B., Cowan T. E., *et al.* Energetic proton generation in ultra intense laser-solid interactions [J]. *Physics of Plasmas*, 2001, 8: 542–549.
[2] Hatchett S. P., Brown C. G., Cowan T. E., *et al.* Electron, photon, and ion beams from the relativistic interaction of petawatt laser pulses with solid targets [J]. *Physics of Plasmas*, 2000, 7: 2076–2082.
[3] Pukhov A. Three-dimensional simulations of ion acceleration from a foil irradiated by a short-pulse laser [J]. *Physical Review Letters*, 2001, 86: 3562–3565.
[4] Mora P. Plasma expansion into a vacuum [J]. *Physical Review Letters*, 2003, 90: 185002.
[5] Mora P. Thin-foil expansion into a vacuum [J]. *Physical Review E*, 2005, 72: 056401.
[6] Passoni M., Lontano M. Theory of light-ion acceleration driven by a strong charge separation [J]. *Physical Review Letters*, 2008, 101: 115001.
[7] Tikhonchuk V. T., Andreev A. A., Bochkarev S. G., Bychenkov V. Y. Ion acceleration in short-laser-pulse interaction with solid foils [J]. *Plasma Physics and Controlled Fusion*, 2005, 47: B869–B877.
[8] Wilks S. C., Kruer W. L., Tabak M., Langdon A. B. Absorption of ultra-intense laser pulses [J]. *Phys. Rev. Lett.*, 1992, 69: 1383–1386.
[9] Malka G., Miquel J. L. Experimental confirmation of ponderomotive-force electrons produced by an ultrarelativistic laser pulse on a solid target [J]. *Physical Review Letters*, 1996, 77: 75–78.
[10] Maksimchuk A., Gu S., Flippo K., *et al.* Forward ion acceleration in thin films driven by a high-intensity laser [J]. *Physical Review Letters*, 2000, 84: 4108–4111.

[11] Clark E. L. Measurements of energetic proton transport through magnetized plasma from intense laser interactions with solids [J]. *Physical Review Letters*, 2000, 84: 670.

[12] Snavely R. A., Key M. H., Hatchett S. P., *et al.* Intense high-energy proton beams from petawatt-laser irradiation of solids [J]. *Physical Review Letters*, 2000, 85: 2945–2948.

[13] Clark E. L., Krushelnick K., Zepf M., *et al.* Energetic heavy-ion and proton generation from ultraintense laser-plasma interactions with solids [J]. *Physical Review Letters*, 2000, 85: 1654–1657.

[14] Mackinnon A. J., Borghesi M., Hatchett S., *et al.* Effect of plasma scale length on multi-MeV proton production by intense laser pulses [J]. *Physical Review Letters*, 2001, 86: 1769–1772.

[15] Allen M., Patel P. K., Mackinnon A., *et al.* Direct experimental evidence of back-surface ion acceleration from laser-irradiated gold foils [J]. *Physical Review Letters*, 2004, 93: 265004.

[16] Fuchs J., Sentoku Y., Karsch S., *et al.* Comparison of laser ion acceleration from the front and rear surfaces of thin foils [J]. *Physical Review Letters*, 2005, 94: 045004.

[17] Schwoerer H., Pfotenhauer S., Jackel O., *et al.* Laser-plasma acceleration of quasi monoenergetic protons from microstructured targets [J]. *Nature*, 2006, 439: 445–448.

[18] Hegelich B. M., Albright B., Cobble J., *et al.* Laser acceleration of quasi-monoenergetic MeV ion beams [J]. *Nature*, 2006, 439: 441–444.

[19] Margarone D., Klimo O., Kim I. J., *et al.* Laser-driven proton acceleration enhancement by nanostructured foils [J]. *Physical Review Letters*, 2012, 109: 234801.

[20] Margarone D., Kim I. J., Psikal J., *et al.* Laser-driven high-energy proton beam with homogeneous spatial profile from a nanosphere target [J]. *Physical Review Special Topics - Accelerators and Beams*, 2015, 18: 071304.

[21] Passoni M., Sgattoni A., Prencipe I., *et al.* Toward high-energy laser-driven ion beams: nanostructured double-layer targets [J]. *Physical Review Accelerators and Beams*, 2016, 19: 061301.

[22] Wagner F., Deppert O., Brabetz C., *et al.* Maximum proton energy above 85 MeV from the relativistic interaction of laser pulses with micrometer thick CH2 targets [J]. *Physical Review Letters*, 2016, 116: 205002.

[23] Esirkepov T., Borghesi M., Bulanov S. V., *et al.* Highly efficient relativistic-ion gener ation in the laser-piston regime [J]. *Physical Review Letters*, 2004, 92: 175003.

[24] Qiao B., Zepf M., Borghesi M., Geissler M. Stable GeV ion-beam acceleration from thin foils by circularly polarized laser pulses [J]. *Physical Review Letters*, 2009, 102: 145002.

[25] Macchi A., Cattani F., Liseykina T. V., Cornolti F. Laser acceleration of ion bunches at the front surface of overdense plasmas [J]. *Physical Review Letters*, 2005, 94: 165003.

[26] Robinson A. P. L., Zepf M., Kar S., *et al.* Radiation pressure acceleration of thin foils with circularly polarized laser pulses [J]. *New Journal of Physics*, 2008, 10: 013021.

[27] Yan X. Q., Lin C., Sheng Z. M., *et al.* Generating high-current monoenergetic proton beams by a circularly polarized laser pulse in the phase-stable acceleration regime [J]. *Physical Review Letters*, 2008, 100: 135003.
[28] Bulanov S. V., Esirkepov T. Z., Koga J., Tajima T. Interaction of electromagnetic waves with plasma in the radiation-dominated regime [J]. *Plasma Physics Reports*, 2004, 30: 196–213.
[29] Bulanov S., Esirkepov T., Migliozzi P., *et al.* Neutrino oscillation studies with laser driven beam dump facilities [J]. *Nuclear Instruments and Methods in Physics Research*, Section A, 2005, 540: 25–41.
[30] Klimo O., Psikal J., Limpouch J., Tikhonchuk V. T. Monoenergetic ion beams from ultrathin foils irradiated by ultrahigh-contrast circularly polarized laser pulses [J]. *Physical Review Special Topics - Accelerators and Beams*, 2008, 11: 031301.
[31] Liseykina T. V., Borghesi M., Macchi A., Tuveri S. Radiation pressure acceleration by ultraintense laser pulses [J]. *Plasma Physics and Controlled Fusion*, 2008, 50: 124033.
[32] Pegoraro F., Bulanov S. V. Photon bubbles and ion acceleration in a plasma dominated by the radiation pressure of an electromagnetic pulse [J]. *Physical Review Letters*, 2007, 99: 065002.
[33] Yan X. Q., Wu H. C., Sheng Z. M., *et al.* Self-organizing GeV, nanocoulomb, collimated proton beam from laser foil interaction at 7×10^{21} W/cm^2 [J]. *Physical Review Letters*, 2009, 103: 135001.
[34] Chen M., Pukhov A., Yu T. P., Sheng Z. M. Enhanced collimated GeV monoenergeticion acceleration from a shaped foil target irradiated by a circularly polarized laser pulse [J]. *Physical Review Letters*, 2009, 103: 024801.
[35] Macchi A., Veghini S., Pegoraro F. "Light sail" acceleration reexamined [J]. *Physical Review Letters*, 2009, 103: 085003.
[36] Yu T. P., Pukhov A., Shvets G., Chen M. Stable laser-driven proton beam acceleration from a two-ion-species ultrathin foil [J]. *Physical Review Letters*, 2010, 105: 065002.
[37] Henig A., Steinke S., Schnurer M., *et al.* Radiation-pressure acceleration of ion beams driven by circularly polarized laser pulses [J]. *Physical Review Letters*, 2009, 103: 245003.
[38] Steinke S., Henig A., Schnurer M., *et al.* Efficient ion acceleration by collective laser driven electron dynamics with ultra-thin foil targets [J]. *Laser and Particle Beams*, 2010, 28: 215–221.
[39] Henig A., Kiefer D., Markey K., *et al.* Enhanced laser-driven ion acceleration in the relativistic transparency regime [J]. *Physical Review Letters*, 2009, 103: 045002.
[40] Kar S., Kakolee K. F., Qiao B., *et al.* Ion acceleration in multispecies targets driven by intense laser radiation pressure [J]. *Physical Review Letters*, 2012, 109: 185006.
[41] Steinke S., Hilz P., Schnurer M., *et al.* Stable laser-ion acceleration in the light sail regime [J]. *Physical Review Special Topics - Accelerators and Beams*, 2013, 16: 011303.
[42] Wang H. Y., Lin C., Sheng Z. M., *et al.* Laser shaping of a relativistic intense,

short gaussian pulse by a plasma lens [J]. *Physical Review Letters*, 2011, 107: 265002.

[43] Bin J. H., Ma W. J., Wang H. Y., *et al.* Ion acceleration using relativistic pulse shaping in near-critical-density plasmas [J]. *Physical Review Letters*, 2015, 115: 064801.

[44] Ma W. J., Kim I. J., Yu J. Q., *et al.* Laser acceleration of highly energetic carbon ions using a double-layer target composed of slightly underdense plasma and ultrathin foil [J]. *Physical Review Letters*, 2019, 122: 014803.

[45] Kim I. J., Pae K. H., Choi I. W., *et al.* Radiation pressure acceleration of protons to 93 MeV with circularly polarized petawatt laser pulses [J]. *Physics of Plasmas*, 2016, 23: 070701.

[46] Higginson A., Gray R. J., King M., *et al.* Near-100 MeV protons via a laser-driven transparency-enhanced hybrid acceleration scheme [J]. *Nature Communications*, 2018, 9: 724.

[47] Silva L. O., Marti M., Davies J. R., *et al.* Proton shock acceleration in laser-plasma interactions [J]. *Physical Review Letters*, 2004, 92: 015002.

[48] Ji L. L., Shen B. F., Zhang X. M., *et al.* Generating monoenergetic heavy-ion bunches with laser-induced electrostatic shocks [J]. *Physical Review Letters*, 2008, 101: 164802.

[49] Fiúza F., Stockem A., Boella E., *et al.* Laser-driven shock acceleration of monoenergetic ion beams [J]. *Physical Review Letters*, 2012, 109: 215001.

[50] Schlegel T., Naumova N., Tikhonchuk V. T., *et al.* Relativistic laser piston model: ponderomotive ion acceleration in dense plasmas using ultraintense laser pulses [J]. *Physics of Plasmas*, 2009, 16: 083103.

[51] Forslund D. W., Shonk C. R. Formation and structure of electrostatic collisionless shocks [J]. *Physical Review Letters*, 1970, 25: 1699.

[52] Haberberger D., Tochitsky S., Fiuza F., *et al.* Collisionless shocks in laser-produced plasma generate monoenergetic high-energy proton beams [J]. *Nature Physics*, 2012, 8: 95–99.

[53] Zhang H., Shen B. F., Wang W. P., *et al.* Collisionless shocks driven by 800 nm laser pulses generate high-energy carbon ions [J]. *Physics of Plasmas*, 2015, 22: 013113.

[54] Yin L., Albright B. J., Hegelich B. M., *et al.* Monoenergetic and GeV ion acceleration from the laser breakout afterburner using ultrathin targets [J]. *Physics of Plasmas*, 2007, 14: 056706.

[55] Yin L., Albright B. J., Bowers K. J., *et al.* Three-dimensional dynamics of breakout afterburner ion acceleration using high-contrast short-pulse laser *and nanoscale targets [J]. Physical Review Letters*, 2011, 107: 045003.

[56] Hegelich B. M., Pomerantz I., Yin L., *et al.* Laser-driven ion acceleration from relativistically transparent nanotargets [J]. *New Journal of Physics*, 2013, 15: 085015.

[57] Wagner F., Bedacht S., Bagnoud V., *et al.* Simultaneous observation of angularly separated laser-driven proton beams accelerated via two different mechanisms [J]. *Physics of Plasmas*, 2015, 22: 063110.

[58] Ditmire T., Tisch J. W. G., Springate E., *et al.* High-energy ions produced in explosions of superheated atomic clusters [J]. *Nature*, 1997, 386: 54–56.

[59] Nishihara K., Amitani H., Murakami M., *et al.* High energy ions generated by laser driven Coulomb explosion of cluster [J]. *Nuclear Instruments and Methods in Physics Research*, Section A, 2001, 464: 98–102.
[60] Fourkal E., Velchev I., Ma C.-M. Coulomb explosion effect and the maximum energy of protons accelerated by high-power lasers [J]. *Physical Review E*, 2005, 71: 036412.
[61] Ostermayr T. M., Haffa D., Hilz P., *et al.* Proton acceleration by irradiation of isolated spheres with an intense laser pulse [J]. *Physical Review E*, 2016, 94: 033208.
[62] Sokollik T., Paasch-Colberg T., Gorling K., *et al.* Laser-driven ion acceleration using isolated mass-limited spheres [J]. *New Journal of Physics*, 2010, 12: 113013.
[63] Pfotenhauer S. M., Jackel O., Polz J., *et al.* A cascaded laser acceleration scheme for the generation of spectrally controlled proton beams [J]. *New Journal of Physics*, 2010, 12: 103009.
[64] Toncian T., Borghesi M., Fuchs J., *et al.* Ultrafast laser-driven microlens to focus and energy-select mega-electron volt protons [J]. *Science*, 2006, 312: 410–413.
[65] Wang H. C., Weng S. M., Murakami M., *et al.* Cascaded acceleration of proton beams in ultrashort laser-irradiated microtubes [J]. *Physics of Plasmas*, 2017, 24: 093117.
[66] He H., Qiao B., Shen X. F., *et al.* All-optical cascaded ion acceleration in segmented tubes driven by multiple independent laser pulses [J]. *Plasma Physics and Controlled Fusion*, 2019, 61: 115005.
[67] Kar S., Ahmed H., Prasad R., *et al.* Guided post-acceleration of laser-driven ions by a miniature modular structure [J]. *Nature Communications*, 2016, 7: 1–7.
[68] Ahmed H., Hadjisolomou P., Naughton K., *et al.* High energy implementation of coil target scheme for guided re-acceleration of laser-driven protons [J]. *Scientific Reports*, 2021,11: 699.
[69] Daido H., Nishiuchi M., Pirozhkov A. S. Review of laser-driven ion sources and their applications [J]. *Reports on Progress in Physics*, 2012, 75: 056401.
[70] Chen Jia'er. Fundamentals of Accelerator Physics [M]. Beijing: Peking University Press, 2012.

Chapter 5

Key Issues in Radiation Pressure Acceleration

From Chapter 4, we know that radiation pressure acceleration (RPA) is a highly promising laser-ion acceleration mechanism, characterized by its good monoenergeticity of the ion beam and high energy conversion efficiency. In the one-dimensional case, the ultrathin target acts as an integral "plasma sail" that is continuously pushed by radiation pressure, allowing ions to be accelerated for an extended period. The faster the "plasma sail" moves, the higher the energy conversion efficiency, which can theoretically approach 100%. Despite these advantages, this mechanism also imposes strict requirements on the acceleration conditions. This chapter will specifically introduce several key issues that currently need to be addressed in RPA, followed by some potentially optimized solutions proposed through theoretical and numerical simulations.

5.1 Realization of Ultra-High Laser Contrast

An ideal laser is a perfect Gaussian pulse without any noise. However, the high-power lasers currently in use are far from this ideal state. In laser-driven ion acceleration experiments, chirped pulse amplification (CPA) technology is commonly used to generate intense laser pulses. Due to limitations in current technology, a series of pre-pulses of varying intensities cannot be completely eliminated before the main pulse arrives. The typical temporal distribution of a laser pulse is shown in Figure 5.1. The laser before the femtosecond main pulse can be collectively referred to as the pulse front or pre-pulse. According to the generation mechanism, from front to back, they can be divided into: (1) Replica pulse: located a few ns to tens of ns before the main pulse, it is a coherent fs-scale short pulse. Some literature also refers to this part as the narrow sense "pre-pulse." This pre-pulse

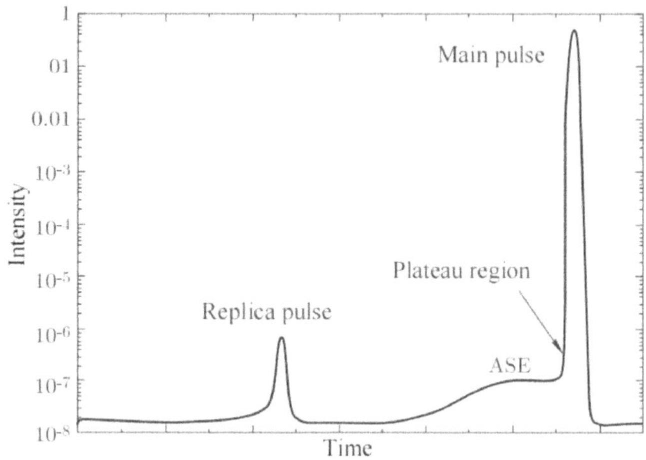

Fig. 5.1 Schematic diagram of the pre-pulse: Before the main laser pulse, there are replication pulses (replica), spontaneous amplification radiation (ASE) and platform pedestal in chronological order, corresponding to the duration of fs, ns and ps magnitude, the intensity depends on the specific parameters of the laser.

is generated due to Fresnel reflection from the surface of the regenerative amplifier crystal, or limited extinction ratio of the beam splitter structure, or insufficient response speed of the Pockels cell to the high-voltage electric pulse, resulting in partial loss of the main pulse. After this pulse is generated, it is output from the regenerative cavity and further amplified in the main amplifier, finally becoming a pulse with a pulse width similar to the main laser after passing through the compressor. (2) Amplified spontaneous emission (ASE): located a few ns before the main pulse, with a pulse duration of ns scale, it is incoherent light. ASE mainly originates from spontaneous emission within the pre-amplifier and main amplifier crystals. Due to the high gain of the Ti:sapphire crystal, the spontaneous emission generated in the gain medium is also amplified in the amplification chain, forming amplified spontaneous emission. (3) Pedestal: generally refers to the region within 100 ps before the main pulse, mainly resulting from the complex dynamical evolution of light within the oscillator, forming a base that extends for about 10 ps in the exponential rise of the main pulse front. It should be noted that Figure 5.1 assumes that the main pulse is close to the Fourier transform limit, so it does not show the fs segment pre-pulse situation due to insufficient dispersion compensation.

In practice, pre-pulses begin to interact with the target before the main pulse arrives, ionizing the target and causing it to expand. The expansion of the target caused by pre-pulses can be divided into two types: front-side expansion and rear-side expansion, which form pre-plasmas with densities decreasing exponentially with length on the front and rear sides of the target, respectively. The front-side pre-plasma is generated by the interaction of pre-pulses with the front surface of the target, heating electrons and causing expansion at the ion acoustic speed: $c_s = \sqrt{Zk_BT/M}$ where k_BT is the electron temperature, Z and M are the ion's charge and mass, respectively. Assuming the pre-plasma is heated to a temperature of $k_BT = 1$ keV, with $Z = 1$ and $M = 1836$, the ion acoustic speed can be estimated as $c_s = 3 \times 10^2$ μm/ns. Rear-side expansion is caused by the shock wave generated by the laser on the front side of the target, which propagates to the rear side and causes plasma expansion. The shock wave speed vs is generally on the order of μm/ns under typical laser and solid target parameters. Since the target has a certain thickness, it takes some time for the shock wave to propagate to the rear side. If the pre-pulse arrives τ laser cycles before the main pulse and the target thickness is d, then the condition for the formation of rear-side pre-plasma is $\tau > d/v_s$. The shock wave speed v_s is positively correlated with the intensity of the pre-pulse; therefore, the weaker the pre-pulse, the less likely it is to form a pre-plasma on the rear side.

The presence of pre-pulses has a non-negligible impact on laser ion acceleration. In experiments dominated by the target normal sheath acceleration (TNSA) mechanism, pre-pulses acting on μm-scale solid targets form pre-plasmas that increase the energy absorption rate of the main pulse, thereby enhancing ion energy to some extent [1–3]. However, too strong pre-pulses can also increase the scale length of the rear-side pre-plasma and reduce the sheath field acceleration gradient [4], which is detrimental to ion energy enhancement. For RPA-dominated acceleration processes, ultrathin targets with nm-scale thickness are typically used, and pre-pulses can easily cause target pre-expansion, disrupting the entire acceleration process [3, 5–7].

To increase ion energy, laser intensity has been continuously improved. Over the past 20 years, an increasing number of PW-class lasers have been constructed and utilized worldwide. Achieving ultra-high laser contrast to reduce or eliminate the negative impact of pre-pulses on ion acceleration is urgently needed. Many techniques have been employed to enhance the temporal contrast of lasers, including optical parametric CPA (OPCPA), cross-polarization (XPW), Pockels cell, saturable absorber, optimized gain crystal pumping, and gratings. These techniques mainly optimize pulses within the laser that are still at lower power. By optimizing the laser, the contrast of femtosecond lasers on the nanosecond scale can reach 10^{11} [11]. For PW lasers, this contrast is still insufficient. Research over the past 20 years has shown that plasma mirror technology is an effective method to improve the laser contrast. Many laboratories around the world have achieved significant results using this method. The specific principle of the plasma mirror will be introduced in Chapter 8.

Currently, many experimental results have been obtained in laser-driven nanotarget ion acceleration, most of which have employed plasma mirror technology. In 2006, Neely *et al.* introduced a single plasma mirror and achieved a proton cutoff energy of 4 MeV [12] using a circularly polarized laser with a wavelength of 800 nm, pulse width of 33 fs, energy of 0.3 J, and intensity of 10^{19} W/cm^2 interacting with a 100 nm aluminum target at the Central Laser Facility (CLF) in the UK. In 2007, Antici *et al.* at the Laboratoire pour l'Utilisation des Lasers Intenses (LULI) in France studied the role of plasma mirrors in proton acceleration. Using a double plasma mirror to significantly enhance the laser contrast, they achieved a proton beam output of 7.3 MeV with an ultrathin SiN target [13]. In 2009, Henig *et al.* based on a double plasma mirror, used a p-polarized laser with a wavelength of 1053 nm, pulse width of 700 fs, energy of 40 J, and intensity of 7×10^{19} W/cm^2 to interact with a 30 nm diamond-like carbon (DLC) thin film target at Los Alamos National Laboratory in the USA, obtaining a proton cutoff energy of 35 MeV [14]. In 2016, at the Gwangju Institute of Science and Technology (GIST) in South Korea, Kim *et al.* used an independent double plasma mirror system and achieved a proton beam cutoff energy of 93 MeV through the RPA acceleration mechanism with circularly polarized femtosecond laser [15]. In 2018, Higginson *et al.* used a single mirror in front of the target and achieved a proton beam cutoff energy exceeding 94 MeV through a hybrid acceleration mechanism with sub-picosecond laser [16].

5.2 Finite Spot Size Effect

In the RPA model, the laser radiation pressure pushes out the electrons in the ultrathin target as a whole. The model is based on the quasi-one-dimensional condition being satisfied during the acceleration process. However, in actual laser acceleration experiments, after being focused by an off-axis parabolic mirror (OAP), the laser has a Gaussian distribution with a strong center and weaker periphery. When using a short-focus OAP, the laser spot full width at half maximum (FWHM) is typically only a few μm, or even focused tightly to 1–2 μm, which is much smaller than the lateral scale of the target. At this point, the quasi-one-dimensional approximation no longer holds. Therefore, when an intense laser irradiates an ultrathin target, the target surface is prone to bend, causing the laser to no longer be incident perpendicularly on the wings of the target. The electrons on the wings will be rapidly heated due to mechanisms such as vacuum heating. This is known as the finite spot size effect. As the electron temperature continues to rise, the target material further expands. Due to the relativistic self-transparency effect, the wings of the target no longer reflect the laser but instead allow it to pass through. Typically, when the plasma density is greater than the critical density n_c, the laser will not penetrate. However, considering the relativistic effect, the laser can propagate in a plasma with a density less than γn_c (where γ is the relativistic factor of the electron), which is equivalent to increasing the threshold density of the plasma that the laser can penetrate. As the target surface continues to bend and expand, the laser penetrates and further heats the electrons. The randomly thermally moving electrons will severely disrupt the RPA acceleration structure, causing the energy spectrum of the accelerated ions to no longer be monoenergetic and increasing the divergence angle of the ion beam. In 2012, Dollar *et al.* irradiated SiN targets of different thicknesses with circularly and linearly polarized light, both with a laser intensity of 2×10^{21} W/cm^2 and a spot FWHM of 1.2 μm. In the case of linearly polarized light, changes in target thickness had little effect on proton cutoff energy, and the proton spectrum was exponentially decreasing in both cases. In the case of circularly polarized light, when the target thickness was reduced from 100 nm to 30 nm, the proton cutoff energy increased from 6 MeV to 13 MeV, and the spectrum remained an exponential decay distribution. Information from the electron spectrum indicates that when the target thickness is further reduced, a large number of electrons are heated. Combined with two-dimensional PIC numerical simulations, it can be seen that when the target becomes thinner, target surface bending leads to electron heating, and the acceleration mechanism remains TNSA.

To suppress the target bending deformation caused by the lateral non-uniformity of the laser intensity, Chen *et al.* proposed a protruding target acceleration scheme. As shown in Figure 5.2, the protruding target can compensate for the negative effects brought by the lateral non-uniformity of the Gaussian laser intensity distribution. Numerical simulations show that the protruding target can better maintain the monoenergeticity of

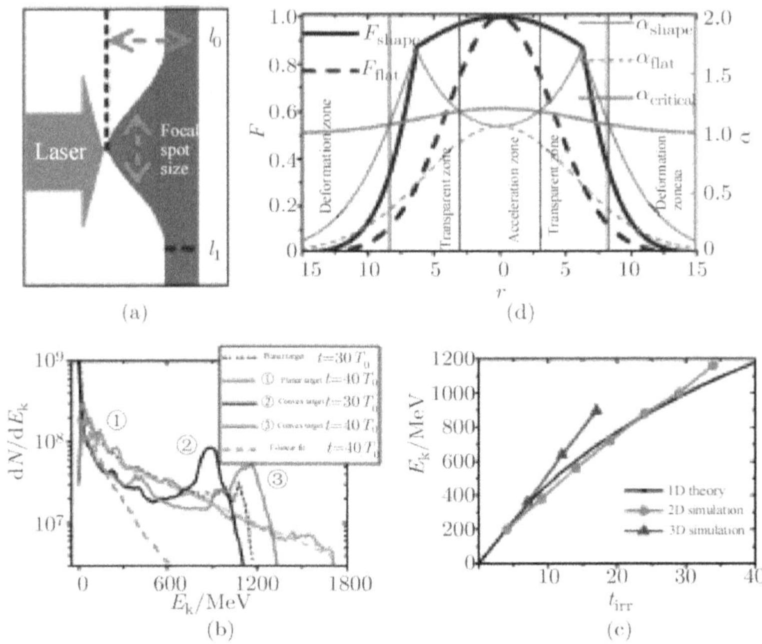

Fig. 5.2 Schematic diagram of the raised target and comparison of acceleration results between the planar target and the raised target [19].

RPA protons, and the divergence angle of the emitted proton beam is also smaller, which is beneficial for the collection and utilization of the proton beam. Subsequently, similar density-modulated target schemes were also proposed. Considering that current target fabrication technology is difficult to achieve such structured targets, and it is also challenging to ensure that the laser is normally incident on the vertex of the protruding target, there is currently a lack of experimental verification of the superiority of this type of target. In addition, some researchers have proposed using a laser with a Hermite–Gaussian distribution in the lateral direction to replace the Gaussian laser to achieve the purpose of suppressing the finite spot size effect.

Furthermore, Xueqing Yan *et al.* found that by adjusting the physical parameters of the interaction between a Gaussian laser and a planar target, the lateral instability that occurs during the acceleration process may achieve ion focusing at the center. This method can theoretically produce a self-focused nC-level proton beam and solve the central breakdown problem. As shown in Figure 5.3, as the acceleration process continues,

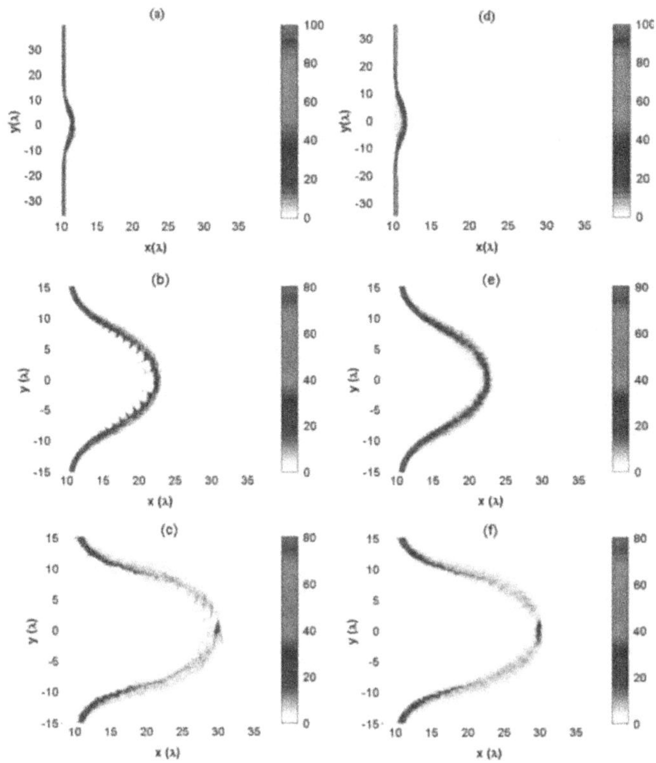

Fig. 5.3 Ion bunching. The first and second columns show the spatial distribution of electrons and protons at three times, respectively [22].

although the electrons and ions on the wings experience density modulation and expansion due to the development of instability, ion focusing occurs at the center.

5.3 Development of Lateral Instability

5.3.1 *Introduction to Lateral Instability*

In numerical simulations of RPA [21, 23] and experiments [24], when the laser radiation pressure pushes the thin target as a whole for acceleration, the target surface will develop ripple structures. These structures continue to evolve, eventually leading to the rupture of the target. The laser will no longer be reflected but will penetrate the target, and the RPA process will terminate prematurely. This process is accompanied by severe electron heating and the disruption of the energy and monoenergeticity of the accelerated ion beam. This phenomenon is known as lateral instability. Lateral instability will occur even when a plane-wave laser with an infinitely large spot interacts with the target. Currently, there are several explanations for the origin of lateral instability in laser ion acceleration, and this section introduces some representative works to help readers understand the current research status of lateral instability.

The mainstream view currently is that lateral instability is a type of Rayleigh–Taylor-like instability (RTI or RT instability). Rayleigh–Taylor instability is a common interface instability in fluids, primarily occurring when a dense heavy fluid is accelerated by a light fluid or when a heavy fluid is placed on top of a light fluid — for example, when water with a higher density is on top of oil with a lower density in a gravitational field. Similarly, the RPA process can be regarded as a light photon fluid pushing a heavy plasma fluid for acceleration, and thus Rayleigh–Taylor instability can also occur. As early as 1972, Ott [26] studied the RT instability and its nonlinear evolution process when a thin layer is pushed by radiation pressure. Later, in 2007, Pegoraro *et al.* further investigated the origin of lateral instability in RPA and considered relativistic effects [27]. The main physical images and conclusions are introduced here, and the specific derivations can be referred to in the original literature. Both Pegoraro and Ott used the Lagrangian method to describe the system, adopted the quasi-neutrality assumption, and treated the target as an infinitely thin single-fluid plane. Initially, the interface is a perfect plane. Subsequently, when the laser radiation pressure begins to act on the single fluid, a small perturbation at the interface will generate a perturbing force in the horizontal direction (perpendicular to the acceleration direction) due to the non-zero gradient of the interface, causing further deformation of the interface. The deformed interface, in turn, increases the horizontal perturbing force. Under the positive feedback effect, the instability continues to develop. Ott further studied the evolution of the interface after entering the nonlinear stage. In the non-relativistic case, both Pegoraro and Ott derived that the growth rate of RTI in the linear stage is $\gamma_{\rm RT} = \sqrt{gk}$, where g is the acceleration and k is the wave number of the instability structure. In the ultra-relativistic case, Pegoraro *et al.* pointed out that the instability growth rate increases with the decrease of laser radiation pressure or the increase of ion mass. This means that the instability growth of light ions will be slower. This is because, at a given laser radiation pressure, light ions can accelerate closer to the speed of light faster than heavy ions, that is, in the laboratory reference frame, time slows down. In the non-relativistic case, the instability grows linearly with time, while in the strong relativistic case, the instability grows with the 1/3 power of time.

In 2012, Palmer *et al.* reported experimental evidence of RT instability in RPA experiments [24]. They used the Vulcan PW laser (with parameters of laser energy 296 ± 7 J, pulse width 660 ± 20 fs, and laser contrast 10^{-10}) to irradiate diamond-like carbon thin targets of 5, 20, 30, and 50 nm thicknesses, and obtained the results shown in Figure 5.4. It can be seen that the proton density exhibits bubble-shaped modulation. The instability modulation wavelength λ_{RT} is near the laser wavelength. However, the RTI growth rate $\gamma_{\mathrm{RT}} = \sqrt{gk}$ (with $k = 2\pi/\lambda_{\mathrm{RT}}$) implies that the smaller the wavelength λ_{RT}, the larger the growth rate, which is inconsistent with the experimental results. The article speculated that this might be due to the scattering and diffraction effects of the laser. Because of scattering and diffraction, the laser radiation pressure is further modulated at the deformed interface, so the growth rate of the instability mode near the laser wavelength is increased. Subsequently, in 2015, Sgattoni *et al.* [28] and Eliasson [29] published articles successively, studying the effect of the lateral modulation of laser radiation pressure caused by interface deformation on the RTI growth mode. After considering this effect, the theoretical model shows an increase in the instability growth rate near the laser wavelength.

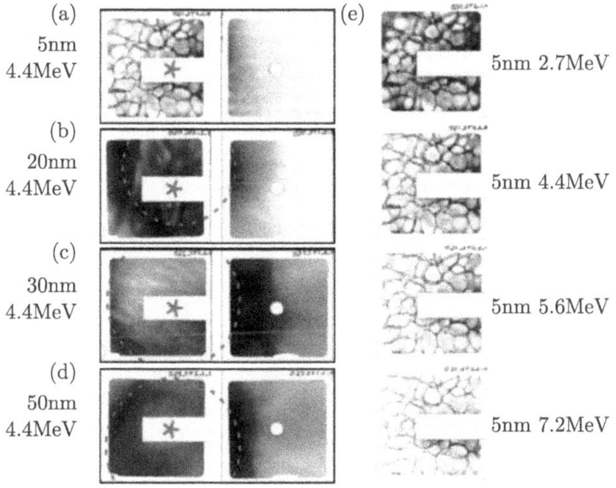

Fig. 5.4 Experimental evidence for RTI: proton distribution [24].

In 2016, Wan *et al.* proposed a new physical mechanism to explain the origin of lateral instability [30]. This mechanism is very similar to the oscillation two-stream instability. As shown in Figure 5.5, when the radiation pressure of circularly polarized laser balances with the electrostatic pressure, a high-density plasma layer is formed. In this thin layer, the laser's lateral electric field oscillates at the laser frequency. A very small ion density fluctuation will couple with the oscillating electric field to excite electron oscillation, which in turn will couple with the laser electric field to produce a laterally non-uniform ponderomotive force, modulate the electron density, further enhance the ion density fluctuation, and then a positive feedback occurs, allowing the instability to develop. The theoretical model predicts the instability structure k_m and growth rate γ_m at maximum growth rate, which are consistent with the results of two-dimensional PIC simulation parameter scans. In 2020, Wan *et al.* presented a theoretical model that simultaneously considered RT instability and the electron-ion coupling effect leading to instability [31]. The theoretical results showed that the growth rate contributed by RT instability is much smaller than that contributed by the electron-ion coupling effect. Therefore, they believe that the electron-ion coupling effect is the dominant mechanism for the origin of lateral instability.

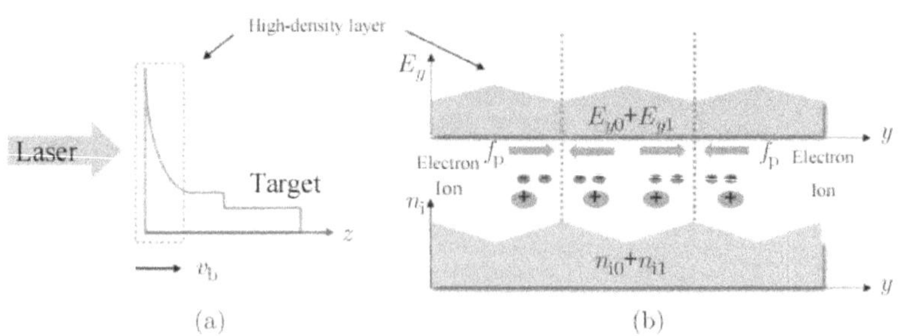

Fig. 5.5 Physical image of excitation instability of electron and ion coupling [30].

In addition to the above viewpoints, there are also views that the origin of lateral instability is a type of Weibel instability [32, 33]. In summary, there is currently no consensus on the physical mechanism of lateral instability development, and further theoretical and experimental exploration is needed.

5.3.2 *Suppression of Lateral Instability*

How to suppress the development of lateral instability is one of the key issues in achieving long-term stable radiation pressure acceleration. In recent years, several schemes discovered through numerical simulations have been proposed successively. Compared to the conventional setup of Gaussian laser and planar thin target, these schemes have all shown varying degrees of instability suppression and enhancement of acceleration effects in the simulation results. On the one hand, many of these schemes are difficult to realize under current experimental conditions. On the other hand, some of them deviate from the physical model of RPA and cannot be considered true RPA. Therefore, there is currently no work that can solve the important problem of lateral instability growth in the RPA process under existing experimental conditions.

This section mainly introduces the above-mentioned optimized schemes for RPA mechanisms. These schemes can be broadly divided into two categories from the perspectives of laser and target configuration. From the target's perspective, there are structured targets, dual-component targets, and heavy ion coatings, etc.; from the laser's perspective, there are elliptically polarized light, multi-pulse laser, single-cycle laser, etc. These will be introduced one by one below.

(a) Structured Target

As shown in Figure 5.6, Chen *et al.* proposed that sub-wavelength structural adjustments on the front surface of a planar target can effectively

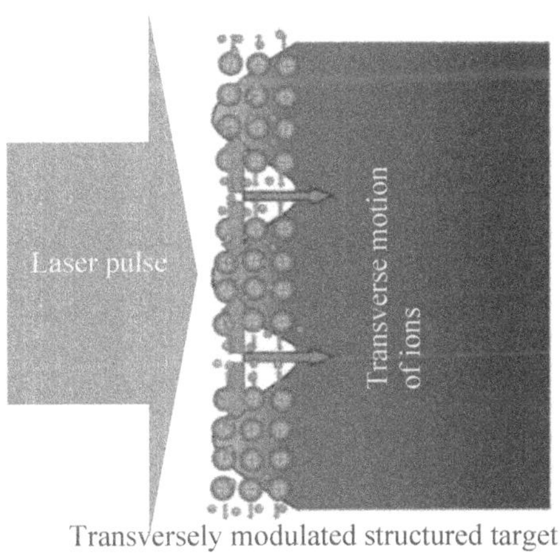

Fig. 5.6 Schematic diagram of structural target suppression instability.

suppress the development of RT instability [34]. When the laser is incident on such a structured target, electrons are pulled out by the lateral oscillating electric field, forming an electrostatic separation field. In this way, ions acquire lateral velocity and begin to diffuse laterally. The lateral diffusion of ions will smooth out modes smaller than the structural wavelength of the target, thereby suppressing instability. The parameter scan of this scheme indicates that when the structural modulation wavelength $\lambda_m = 0.4\lambda_L$ (where λ_L is the laser wavelength), the highest proton cutoff energy and the best monoenergeticity are obtained.

(b) Dual-Component Target

In 2010, Yu *et al.* proposed that a dual-component carbon-hydrogen target can effectively suppress the development of RT instability and achieve good monoenergetic proton acceleration [35]. Because carbon and hydrogen have different charge-to-mass ratios, in the early stage of acceleration, the lighter protons will separate from the heavier carbon ions. As shown in Figure 5.7(a), the carbon ion layer continuously expands during the acceleration process and provides a protection for the proton layer. In this way, the proton layer will not develop ripple structures due to the development of RT instability, preventing the laser from penetrating the target.

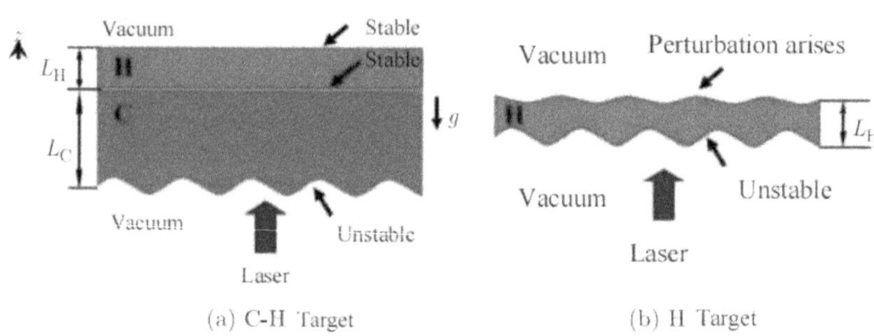

Fig. 5.7 Comparison of acceleration effect between two-component target (a) and single-component target (b) [35].

(c) Dynamic Stabilization Acceleration

The dynamic stabilization acceleration scheme is realized by coating a thin layer of high atomic number material (such as gold) on the front surface of an ultrathin target. This scheme was proposed by Shen *et al.* in 2017. Because gold has a high atomic number and does not fully ionize at the beginning but continuously ionizes to produce electrons, the coated film acts as an "electron reservoir," continuously providing electrons during the

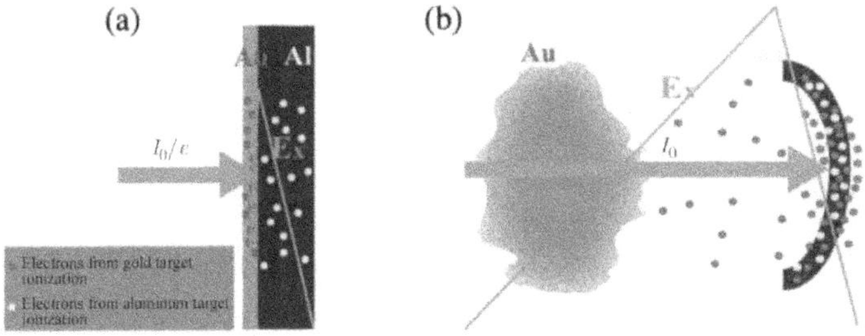

Fig. 5.8 Acceleration diagram of heavy ion coating target. Ex represents the accelerating electric field [36].

sail acceleration process. As shown in Figure 5.8, without the coating, the development of instability will lead to continuous electron heating, causing the destruction of the acceleration structure: using "sail" as a metaphor, the "sail" used for acceleration is broken. The electrons provided by the coating layer are equivalent to continuously repairing the "sail," allowing the acceleration process to be maintained for a longer time, thereby achieving better acceleration effects. Three-dimensional PIC simulation results show that when the laser intensity is 10^{22} W/cm^2, an Al^{13+} ion beam with a charge greater than 20 nC and a cutoff energy reaching 3.8 GeV can be obtained.

(d) Elliptically Polarized Laser

It is generally believed that circularly polarized light is more favorable for RPA because its ponderomotive force has no 2ω oscillation term, which can suppress electron heating. However, Wu *et al.* proposed that in the acceleration process driven by elliptically polarized laser, due to the moderate $\boldsymbol{J} \times \boldsymbol{B}$ effect of elliptically polarized laser heating electrons, ions can obtain a faster lateral diffusion speed, which can suppress RT instability. The lateral diffusion speed of ions can be measured by the ion acoustic speed $(ZT_\mathrm{e}/m_\mathrm{i})^{1/2}$, where T_e is the electron temperature, Z is the ion's charge number, and m_i is the ion mass. To increase the lateral speed of ions, it is necessary to increase the electron temperature, that is, to reduce the laser polarization degree $\alpha = a_z/a_y$ (where $\alpha = 1$ for a circularly polarized light and $\alpha = 0$ for a linearly polarized light). On the other hand, if the polarization degree is too small, the RPA structure will be destroyed. Therefore, for a given laser intensity and plasma parameters, there exists an optimal laser polarization degree. The upper limit of the polarization degree is determined by $\tau v_\mathrm{d} > \lambda$, where v_d is the ion acoustic speed, τ is the characteristic time for RTI growth, and λ is the wavelength of the instability development structure. The lower limit of the polarization degree needs to ensure that the electrons pulled out by the elliptically polarized laser can be pulled back into the dense electron layer by the electrostatic force and laser ponderomotive force. Two-dimensional PIC simulations further confirm the superiority of elliptically polarized light. As shown in Figure 5.9, the three rows from top to bottom are the electron density and ion density distributions for a circularly polarized light ($\alpha = 1$), elliptically polarized light ($\alpha = 0.7$), and elliptically polarized light ($\alpha = 0.3$), respectively. $\alpha = 0.7$ is within the appropriate polarization degree range, while

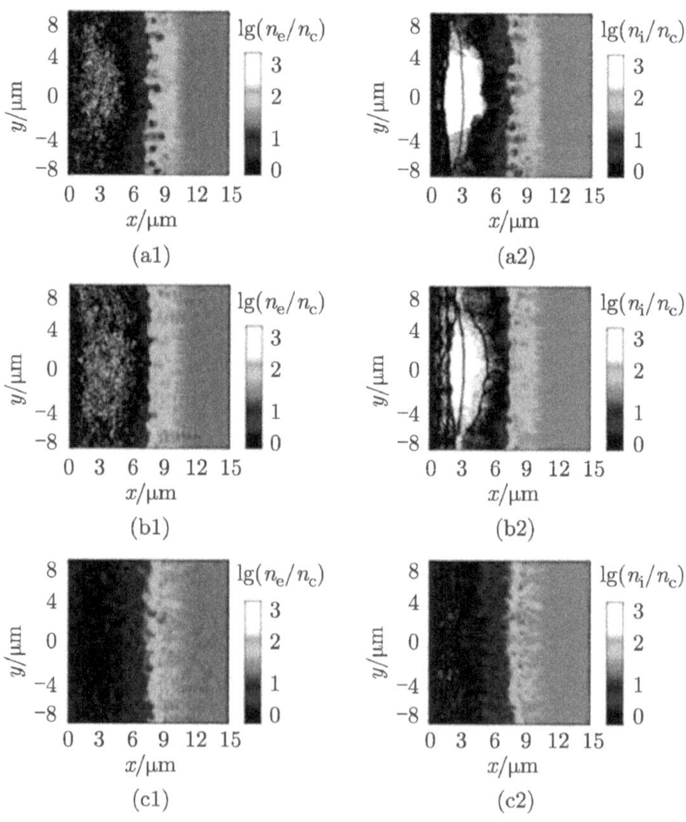

Fig. 5.9 Comparison of the effects of different polarization degree lasers on suppressing instability development. The three rows from top to bottom are the electron density and ion density distributions for a circularly polarized light ($\alpha = 1$), elliptically polarized light ($\alpha = 0.7$), and elliptically polarized light ($\alpha = 0.3$) [37].

$\alpha = 0.3$ is too low. It can be seen from the figure that compared with the cases of $\alpha = 1$ and $\alpha = 0.3$, lateral instability is more effectively suppressed in the case of $\alpha = 0.7$.

(e) Multi-Pulse Laser

The spatial layout of the multi-pulse laser is completely consistent longitudinally and arranged in a circle with a certain radius laterally, achieving an effect of low intensity in the center and high intensity at the edge. This distribution of laser intensity can be realized by the future high-power integrated fiber laser conceived by Professor Mourou [43]. Figure 5.10 shows the lateral intensity distribution of the multi-pulse laser. The multi-pulse laser ion acceleration scheme proposed by Zhou *et al.* introduced the coupling of long-wavelength modes, thereby suppressing the growth of the short-wavelength modes inherent in the target that easily cause rapid target rupture, which is conducive to extending the ion acceleration time and obtaining higher ion energy [38]. In the traditional RPA acceleration mechanism, the perturbation wavelength of RTI is in the same order of magnitude as the laser wavelength λ, which can be regarded as short-wavelength mode perturbation. Short-wavelength perturbations develop very quickly, and generally, within dozens of cycles, the accelerated beam will completely disperse. The multi-pulse acceleration mechanism introduces a concave structure of nearly 20 wavelengths in the laser's lateral direction. Compared with the perturbation of the order of laser wavelength λ, this structure causes a long-wavelength mode perturbation. Theoretical derivation has found that when short-wave and long-wave modes are coupled, the growth of short-wavelength modes is inhibited, and at the same time, the divergence angle of ions decreases with time. The first and second rows in Figure 5.11 are the two-dimensional PIC simulation results of ion density evolution for

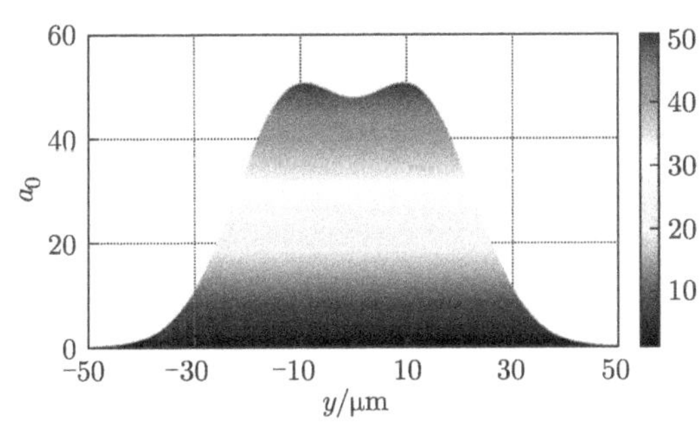

Fig. 5.10 Lateral intensity distribution of multi-pulse laser.

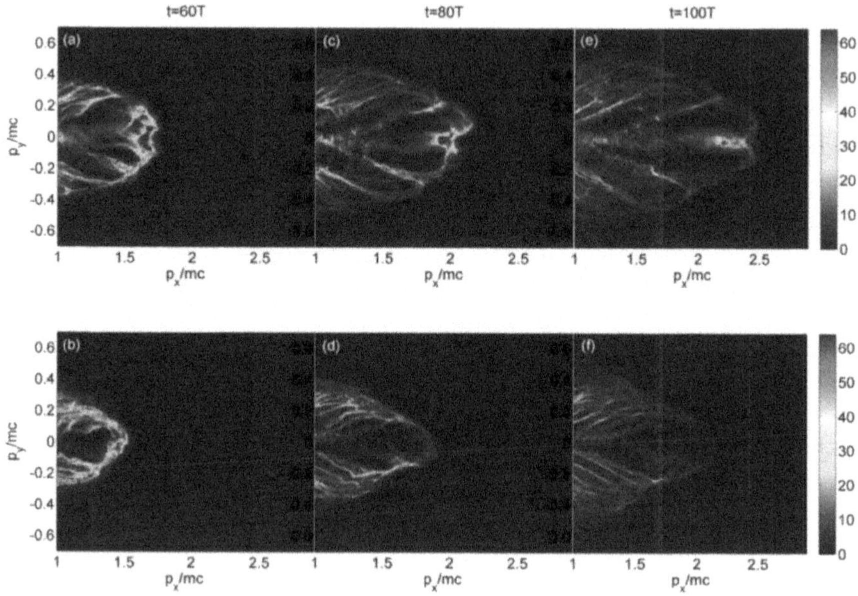

Fig. 5.11 The first and second rows are the two-dimensional PIC simulation results of ion density evolution for multi-pulse and single-pulse laser ion acceleration, respectively.

multi-pulse and single-pulse laser ion acceleration, respectively. It can be seen from the figure that under the drive of a multi-pulse laser, ions can be accelerated for a long time, and the divergence angle continuously decreases, while under the drive of single-pulse laser, ions gradually disperse due to the development of lateral instability.

(f) Single-Cycle Laser

The development of laser technology will continuously promote the development of ion acceleration research. After 2010, Mourou *et al.* proposed the thin film compressor (TFC) technology. This technology can compress laser pulses with energy in the joule level and a pulse duration of tens of femtoseconds to the 1–2 fs level, thereby obtaining ultra-high peak power single-cycle lasers. Simply put, the thin film compressor technology uses the nonlinear optical effect to broaden the spectrum of the laser that has been compressed close to the Fourier limit by CPA technology, enabling it to support shorter pulses, and then compresses it again. Figure 5.12 shows a pulse with a pulse width of 25 fs and energy of 27 J being compressed to 5 fs and finally to 2 fs through dual thin film compression.

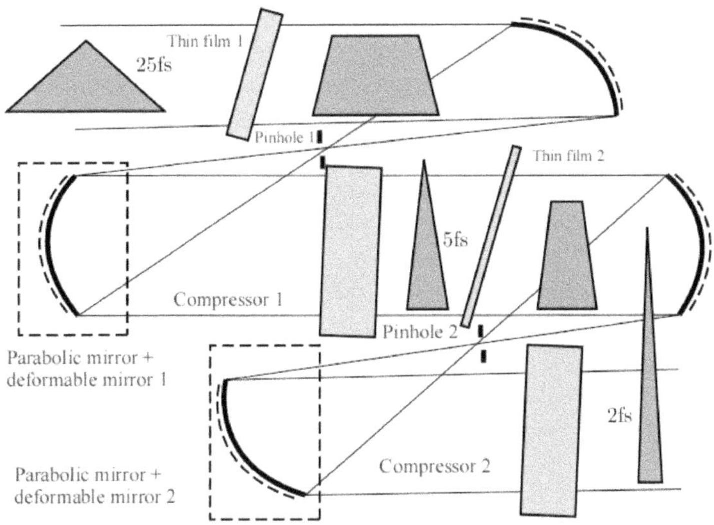

Fig. 5.12 Schematic diagram of dual thin film compression technology [39].

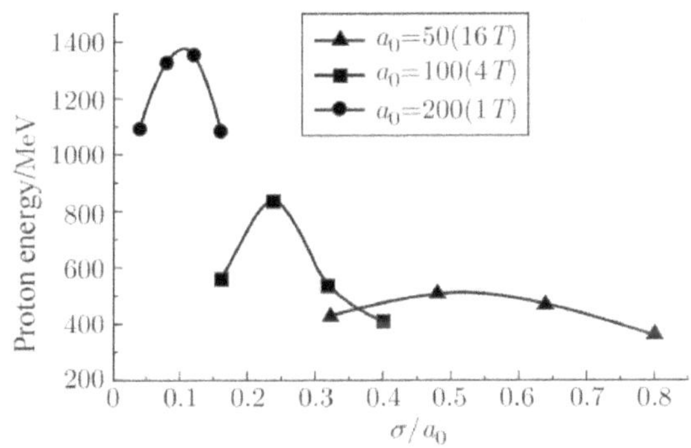

Fig. 5.13 Comparison of proton cutoff energy between single-cycle and multi-cycle laser ion acceleration under the same laser energy [40].

Zhou *et al.* found that proton acceleration driven by ultra-intense single-cycle laser does not develop lateral instability, and compared with multi-cycle laser pulses, it has other advantages [40]. As shown in Figure 5.13, under the same laser energy, single-cycle laser can obtain higher proton cutoff energy than multi-cycle laser. Parameter scanning found that the optimal target surface density parameter in single-cycle laser proton acceleration is different from that in multi-cycle laser. In the traditional multi-cycle laser-driven RPA mechanism, the optimal target surface density satisfies $\sigma_{\text{op}} \sim a_0$, while in the single-cycle laser case, $\sigma_{\text{op}} \sim 0.12a_0$, that is, the laser radiation pressure is much greater than the maximum electrostatic pressure that the target can provide. Therefore, the single-cycle laser ion acceleration mechanism is essentially different from the RPA mechanism. As shown in Figure 5.14, when a single-cycle laser irradiates an ultrathin target, because the laser radiation pressure is much greater than the electrostatic pressure, electrons are pushed out as a whole, thereby forming an electrostatic separation field, and then protons are accelerated and gain energy in the formed electrostatic separation field. Because there is no lateral instability development, such an acceleration structure can be maintained for a long time, so the proton cutoff energy is high. Subsequently, an electrostatic capacitor acceleration mechanism driven by few-cycle laser was proposed. The expulsion of electrons as a whole forms an electrostatic separation field, which can be described as the charging process of a capacitor. The acceleration of protons in the electrostatic field can be understood as the discharge process of the capacitor.

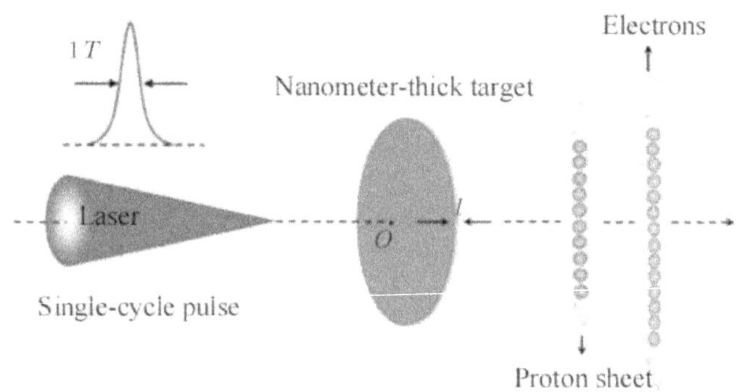

Fig. 5.14 Schematic diagram of single-cycle laser acceleration mechanism [40].

Comparing the two acceleration methods mentioned above, it can be found that the single-cycle laser ion acceleration mechanism is essentially different from the RPA mechanism. In single-cycle laser acceleration, electrons and protons are separated, while in RPA, electrons and ions are a whole, and they are pushed together for acceleration by radiation pressure. On the other hand, from the perspective of energy transfer, there is also a difference in energy conversion efficiency between the two mechanisms. In the ideal RPA acceleration process, laser energy is continuously transferred to the "plasma sail" composed of electrons and ions, and because the mass of ions is much greater than that of electrons, most of this energy is transferred to the ions. As the speed of the "plasma sail" increases, the energy conversion efficiency from laser to ions also increases. In single-cycle laser acceleration, when electrons are pushed out to form an electrostatic field, laser energy is mainly stored in the form of electrostatic energy, and then the ion acceleration process can be regarded as the process of electrostatic energy continuously converting into ion kinetic energy. Overall, laser energy is mainly consumed in the expelled electron sheet. Therefore, from the perspective of energy conversion efficiency, the single-cycle acceleration mechanism does not have an advantage compared to the RPA mechanism. Based on this, Wu *et al.* further found that under the same laser energy conditions, few-cycle laser (pulse greater than single-cycle) can obtain higher laser-to-ion energy conversion efficiency compared to single-cycle and multi-cycle laser [42].

References

[1] Snavely R. A., Key M. H., Hatchett S. P., *et al.* Intense high-energy proton beams from petawatt-laser irradiation of solids [J]. *Physical Review Letters*, 2000, 85(14): 2945–2948.

[2] Mackinnon A. J., Sentoku Y., Patel P. K., *et al.* Enhancement of proton acceleration by hot-electron recirculation in thin foils irradiated by ultraintense laser pulses [J]. *Physical Review Letters*, 2002, 88(21): 215006.

[3] Zepf M., Clark E. L., Beg F. N., *et al.* Proton acceleration from high-intensity laser interactions with thin foil targets [J]. *Physical Review Letters*, 2003, 90(6): 064801.

[4] Lundh O., Lindau F., Persson A., *et al.* Influence of shock waves on laser-driven proton acceleration [J]. *Physical Review E*, 2007, 76(2): 026404.

[5] Esirkepov T., Borghesi M., Bulanov S. V., *et al.* Highly efficient relativistic-ion generation in the laser-piston regime [J]. *Physical Review Letters*, 2004, 92(17): 175003.

[6] Qiao B., Zepf M., Borghesi M., *et al.* Stable GeV ion-beam acceleration from thin foils by circularly polarized laser pulses [J]. *Physical Review Letters*, 2009, 102(14): 145002.

[7] Yan X. Q., Tajima T., Hegelich M., *et al.* Theory of laser ion acceleration from a foil target of nanometer thickness [J]. *Applied Physics B*, 2010, 98(4): 711–721.

[8] Danson C. N., Haefner C., Bromage J., *et al.* Petawatt and exawatt class lasers world wide [J]. *High Power Laser Science and Engineering*, 2019, 7: e54.

[9] Kiriyama H., Shimomura T., Sasao H., *et al.* Temporal contrast enhancement of petawatt-class laser pulses [J]. *Opt. Lett.*, 2012, 37(16): 3363–3365.

[10] Tang Y., Hooker C. J., Chekhlov O. V., *et al.* Novel contrast enhancement of Astra Gemini laser facility [C]. CLEO: 2013. IEEE, 2013: 1–2.
[11] Sung J. H., Lee H. W., Yoo J. Y., *et al.* 4.2 PW, 20 fs Ti: sapphire laser at 0.1 Hz [J]. *Opt. Lett.*, 2017, 42(11): 2058–2061.
[12] Neely D., Foster P., Robinson A., *et al.* Enhanced proton beams from ultrathin targets driven by high contrast laser pulses [J]. *Appl. Phys. Lett.*, 2006, 89(2): 021502.
[13] Antici P., Fuchs J., D'humiéres E., *et al.* Energetic protons generated by ultrahigh contrast laser pulses interacting with ultrathin targets [J]. *Physics of Plasmas*, 2007, 14(3): 030701.
[14] Henig A., Kiefer D., Markey K., *et al.* Enhanced laser-driven ion acceleration in the relativistic transparency Regime [J]. *Physical Review Letters*, 2009, 103(4): 045002.
[15] Kim I. J., Pae K. H., Choi I. W., *et al.* Radiation pressure acceleration of protons to 93 MeV with circularly polarized petawatt laser pulses [J]. *Physics of Plasmas*, 2016,23(7): 070701.
[16] Higginson A., Gray R. J., King M., *et al.* Near-100 MeV protons via a laser-driven transparency-enhanced hybrid acceleration scheme [J]. *Nature Communications*, 2018, 9(1): 724.
[17] Dollar F., Zulick C., Thomas A., *et al.* Finite spot effects on radiation pressure acceleration from intense high-contrast laser interactions with thin targets [J]. *Physical Review Letters*, 2012, 108(17): 175005.
[18] Vshivkov V. A., Naumova N. M., Pegoraro F., *et al.* Nonlinear electrodynamics of the interaction of ultra-intense laser pulses with a thin foil [J]. *Physics of Plasmas*, 1998, 5(7): 2727–2741.
[19] Chen M., Pukhov A., Yu T., *et al.* Enhanced collimated GeV monoenergetic ion acceleration from a shaped foil target irradiated by a circularly polarized laser pulse [J]. *Physical Review Letters*, 2009, 103(2): 024801.
[20] Yu T. P., Chen M., Pukhov A. High quality GeV proton beams from a density modulated foil target [J]. *Laser Part Beams*, 2009, 27(4): 611–617.
[21] Chen M., Pukhov A., Sheng Z., *et al.* Laser mode effects on the ion acceleration during circularly polarized laser pulse interaction with foil targets [J]. *Physics of Plasmas*, 2008, 15(11): 113103.
[22] Yan X. Q., Wu H. C., Sheng Z. M., *et al.* Self-organizing GeV, nanocoulomb, collimated proton beam from laser foil interaction at 7×10^{21} W/cm^2 [J]. *Physical Review Letters*, 2009, 103(13): 135001.
[23] Robinson A., Zepf M., Kar S., *et al.* Radiation pressure acceleration of thin foils with circularly polarized laser pulses [J]. *New journal of Physics*, 2008, 10(1): 013021.
[24] Palmer C., Schreiber J., Nagel S., *et al.* Rayleigh–Taylor instability of an ultrathin foil accelerated by the radiation pressure of an intense laser [J]. *Physical Review Letters*, 2012, 108(22): 225002.
[25] Paradkar B., Krishnagopal S. Electron heating in radiation-pressure-driven proton acceleration with a circularly polarized laser [J]. *Physical Review E*, 2016, 93(2): 023203.
[26] Ott E. Nonlinear evolution of the Rayleigh–Taylor instability of a thin layer [J]. *Physical Review Letters*, 1972, 29(21): 1429.

[27] Pegoraro F., Bulanov S. V. Photon bubbles and ion acceleration in a plasma dominated by the radiation pressure of an electromagnetic pulse [J]. *Physical Review Letters*, 2007, 99(6): 065002.
[28] Sgattoni A., Sinigardi S., Fedeli L., *et al.* Laser-driven Rayleigh–Taylor instability: Plasmonic effects and three-dimensional structures [J]. *Physical Review E*, 2015, 91(1): 013106.
[29] Eliasson B. Instability of a thin conducting foil accelerated by a finite wavelength intense laser [J]. *New Journal of Physics*, 2015, 17(3): 033026.
[30] Wan Y., Pai C.-H., Zhang C., *et al.* Physical mechanism of the transverse instability in radiation pressure ion acceleration [J]. *Physical Review Letters*, 2016, 117(23): 234801.
[31] Wan Y., Andriyash I., Lu W., *et al.* Effects of the transverse instability and wave breaking on the laser-driven thin foil acceleration [J]. *Physical Review Letters*, 2020, 125(10): 104801.
[32] Zhang X., Shen B., Ji L., *et al.* Instabilities in interaction of circularly polarized laser pulse and overdense target [J]. *Physics of Plasmas*, 2011, 18(7): 073101.
[33] Yan X., Chen M., Sheng Z., *et al.* Self-induced magnetic focusing of proton beams by Weibel-like instability in the laser foil-plasma interactions [J]. *Physics of Plasmas*, 2009, 16(4): 044501.
[34] Chen M., Kumar N., Pukhov A., *et al.* Stabilized radiation pressure dominated ion acceleration from surface modulated thin-foil targets [J]. *Physics of Plasmas*, 2011, 18(7): 073106.
[35] Yu T.-P., Pukhov A., Shvets G., *et al.* Stable laser-driven proton beam acceleration from a two-ion-species ultrathin foil [J]. *Physical Review Letters*, 2010, 105(6): 065002.
[36] Shen X., Qiao B., Zhang H., *et al.* Achieving stable radiation pressure acceleration of heavy ions via successive electron replenishment from ionization of a high-z material coating [J]. *Physical Review Letters*, 2017, 118(20): 204802.
[37] Wu D., Zheng C., Qiao B., *et al.* Suppression of transverse ablative Rayleigh–Taylor like instability in the hole-boring radiation pressure acceleration by using elliptically polarized laser pulses [J]. *Physical Review E*, 2014, 90(2): 023101.
[38] Zhou M., Liu B., Hu R., *et al.* Stable radiation pressure acceleration of ions by suppress ing transverse Rayleigh–Taylor instability with multiple Gaussian pulses [J]. *Physics of Plasmas*, 2016, 23(8): 083109.
[39] Mourou G., Mironov S., Khazanov E., *et al.* Single cycle thin film compressor opening the door to zeptosecond-exawatt physics [J]. *The European Physical Journal Special Topics*, 2014, 223(6): 1181–1188.
[40] Zhou M., Yan X., Mourou G., *et al.* Proton acceleration by single-cycle laser pulses offers a novel monoenergetic and stable operating regime [J]. *Physics of Plasmas*, 2016, 23(4): 043112.
[41] Shen X. F., Qiao B., Zhang H., *et al.* Electrostatic capacitance-type acceleration of ions with an intense few-cycle laser pulse [J]. *Appl. Phys. Lett.*, 2019, 114(14): 144102.

[42] Wu X. Z., Gong Z., Shou Y. R., *et al.* Efficiency enhancement of ion acceleration from thin target irradiated by multi-PW few-cycle laser pulses [J]. *Physics of Plasmas*, 2021, 28(2): 023102.

[43] Mourou G. A., Tajima T., Bulanov S. V. Optics in the relativistic regime [J]. *Rev. Mod. Phys.*, 2006, 78: 309

Chapter 6

Introduction to Ti:Sapphire Ultrashort and Ultrastrong Lasers

In 1960, the first laser was born [1]. Since then, people have continuously developed new laser technologies, pursuing higher laser energies to expand the scope and applications of lasers. In 1962, the Q-switching technique was proposed [2]. By modulating the loss of the optical resonator, it achieved a sudden change in the Q-value of the optical resonator, enabling the output of giant laser pulses. The Q-switching technique first reduced the laser pulse width to the nanosecond (ns) level, with the corresponding laser peak power reaching the megawatt (MW, 10^6 W) level. In 1964, the mode-locking technique emerged [3]. By periodically modulating the laser pulses within the resonator, it ensured a constant phase difference between the longitudinal modes of the cavity, thereby forming a sequence of ultrashort pulses with equal time intervals. The mode-locking technique further shortened the laser pulse width to the picosecond (ps) level, and the corresponding peak power increased to the gigawatt (GW, 10^9 W) level. In 1981, mode-locking was applied to dye lasers, achieving ultrashort laser pulses of 90 fs [4]. This important breakthrough brought the laser pulse width to the femtosecond (fs) level, enabling a wide range of applications. With the advent of Ti:sapphire laser technology in 1982 [5], and the realization of 60-fs ultrashort laser pulses using Kerr-lens mode-locking in 1991 [6], commercial products with sub-10-fs levels have been achieved today. Ti:sapphire lasers, with their simple structure, stable performance, and widely tunable output wavelengths, have gradually become the mainstream technology in the field of femtosecond lasers. Although these lasers have successfully compressed the pulse width to the fs level, they still face the limitation of insufficient peak power; this is because once the peak power exceeds a certain threshold, the laser crystal will suffer permanent damage due to nonlinear effects. In 1985, Mourou *et al.* proposed the chirped pulse amplification (CPA) technique, which effectively solved the problem of optical element damage, allowing laser output power to reach the terawatt (TW, 10^{12} W) and even petawatt (PW, 10^{15} W) levels. This chapter will introduce ultrashort and ultrastrong femtosecond laser systems based on Ti:sapphire gain media.

6.1 Ti:Sapphire Oscillator

A femtosecond Ti:sapphire oscillator is a laser that uses Ti-doped sapphire as the gain medium to produce femtosecond laser pulses. Like ordinary lasers, a femtosecond Ti:sapphire oscillator generally consists of three basic elements: a pump source, a gain medium, and a resonator. The basic structure of a Ti:sapphire oscillator is shown in Figure 6.1. The gain medium, Ti-doped sapphire (Ti:S), is cut at the Brewster angle and is located in the middle of a symmetrical confocal structure formed by two concave mirrors with equal radii of curvature, M2 and M3 (M3 being a dichroic mirror). A high reflectivity mirror, M1, and a coupling output mirror, M4, form the resonator. The pump light is focused onto the Ti-doped sapphire via lens L1. P1 and P2 are a pair of prisms used to regulate the dispersion within the cavity. Unlike ordinary lasers, femtosecond Ti:sapphire lasers have special requirements for the fluorescence spectral bandwidth of the gain medium, mode-locking conditions, and dispersion control. These three issues will be discussed in detail in this section.

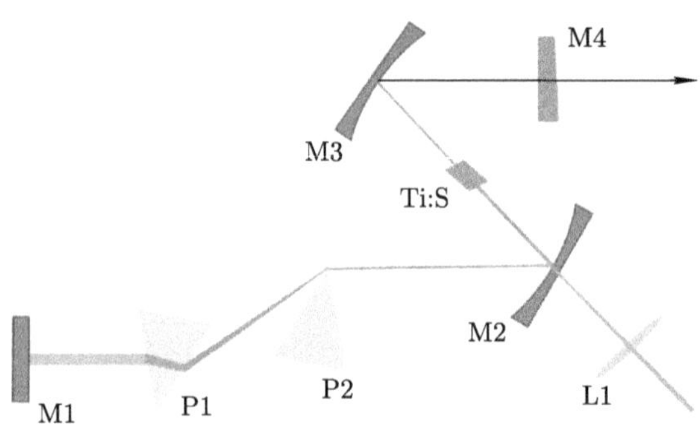

Fig. 6.1 Schematic diagram of a basic structure of a femtosecond Ti:sapphire oscillator.

6.1.1 *Ti:Sapphire Crystal*

Ti:sapphire, also known as Ti-doped sapphire, is a single crystal of Al_2O_3 doped with Ti^{3+} ions, belonging to the hexagonal crystal system. Its

Table 6.1 Physicochemical parameters of Ti:Sapphire crystal [8].

Chemical Formula	Crystal Structure	Lattice Constants	Density	Melting Point	Mohs Hardness	Thermal Conductivity	Specific Heat Capacity
Ti^{3+}: Al_2O_3	Hexagonal system	$a = 4.758$Å* $c = 12.99$Å	3.98 g/cm^3	2040°C	9	52 W/(m·K)	0.42 J/(g·K)
Fluorescence lifetime	**Tuning range**	**Absorption range**	**Emission peak**	**Absorption peak**	**Refractive index**	**Peak cross-section**	**Thermal expansion coefficient**
3.2μs ($T =$ 300 K)	660–1050 nm	400–600 nm	795 nm	488 nm	1.76 (at 800 nm)	(3–4) $\times 10^{-19}$cm^2	8.40 $\times 10^{-6}$/°C

$*$ 1 Å = 10^{-10} m.

physicochemical properties consist of: good stability, a thermal conductivity three times better than that of YAG crystals, and good processability, making it an excellent laser medium material. The representative parameters are shown in Table 6.1.

The absorption and fluorescence spectra of Ti:sapphire crystals are shown in Figure 6.2. The absorption spectrum, depicted in Figure 6.2(a), peaks at approximately 490 nm with a spectral range between 400 and 650 nm. This broad band allows for pumping by various types of lasers,

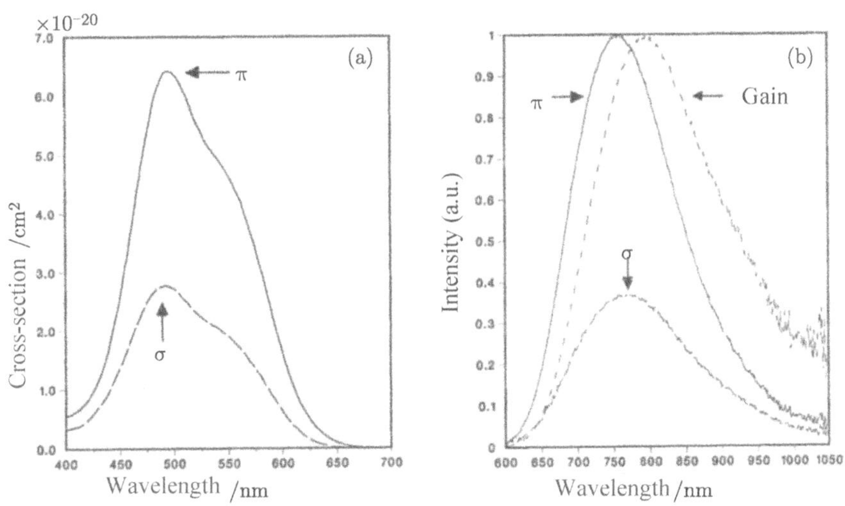

Fig. 6.2 Absorption (a) and fluorescence (b) spectra of Ti-doped sapphire crystals. P. F. Moulton, "Spectroscopic and laser characteristics of Ti:Al_2O_3," J. Opt. Soc. Am. B 3, 125–133 (1986).

such as frequency-doubled Nd:YAG lasers and semiconductor lasers. The fluorescence spectrum, illustrated in Figure 6.2(b), centers around 790 nm with a spectral range extending from 600 to 1200 nm. Considering the relationship between fluorescence intensity and gain coefficient, the gain peak is found near 790–800 nm, with a full width at half maximum (FWHM) of the gain curve exceeding 200 nm. Theoretically, under Fourier-transform-limited conditions, this supports the generation of laser pulses shorter than 3 fs. Additionally, it is evident from Figure 6.2 that both the absorption and fluorescence spectra exhibit polarization selectivity. π-polarization is advantageous for both the transfer of pump laser energy and optimal laser output, hence it is typically employed as the working state.

6.1.2 *Ultrashort Pulse Technology*

In the free-running state, a laser typically oscillates in multiple longitudinal modes, with each mode being independent and having irregular phase and amplitude distributions. Mode-locking technology can correlate these modes, aligning their phases, which is crucial for achieving ultrashort pulse widths through coherent addition of multiple modes, as illustrated in Figure 6.3.

Suppose a laser has $2N+1$ longitudinal modes oscillating. For a multi-longitudinal-mode laser, the electric field intensity of each longitudinal mode is [11]

$$E_q(z,t) = E_q \mathrm{e}^{\mathrm{i}\left[\omega_q\left(t-\frac{z}{v}\right)+\varphi_q\right]}. \tag{6.1.1}$$

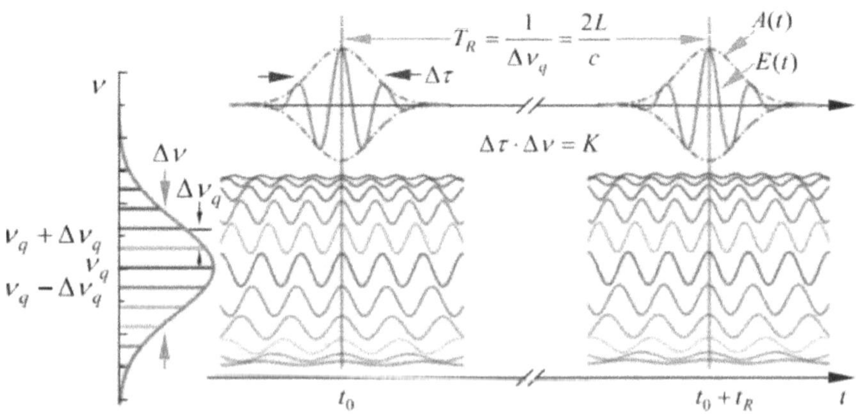

Fig. 6.3 Schematic diagram of mode-locking for ultrashort pulse generation [10].

In the absence of mode-locking, the total electric field $E(t)$ is the sum of the electric fields of the $2N + 1$ modes:

$$E(t) = \sum_{q=-N}^{N} E_q \cos(\omega_q t + \varphi_q), \tag{6.1.2}$$

where ω_q is the angular frequency of the q-th mode, φ_q is the initial phase of the q-th mode, and E_q is the amplitude of the q-th mode. Due to the incoherent addition of the modes, the output light intensity exhibits random, irregular fluctuations, with the average intensity $\overline{I}$ being the sum of the intensities of the individual modes. If the amplitudes of the modes are equal, denoted as E_0, then $I \propto (2N + 1)E_0^2$.

If a method is employed to ensure that the frequency intervals between the oscillating modes are equal and that they possess a definite phase relationship, then the laser can output extremely narrow pulses with high peak power. Assuming that the frequency intervals and phase relationships are as follows:

$$\omega_q = \omega_0 + q\Omega, \tag{6.1.3}$$

$$\varphi_q = \varphi_0 + q\eta, \tag{6.1.4}$$

where Ω is the angular frequency difference between adjacent modes: $\Omega = \dfrac{\pi c}{L'}$, with L' being the cavity length; η is a constant.

At $z = 0$, the electric field of the q-th mode is

$$E_q(t) = E_q \mathrm{e}^{\mathrm{i}[(\omega_0 + q\Omega)t + \varphi_0 + q\eta]}. \tag{6.1.5}$$

Under mode-locking conditions, the total electric field is the coherent sum of the $2N + 1$ modes, assuming each mode has an electric field of E_0:

$$E(t) = E_0 \mathrm{e}^{\mathrm{i}(\omega_0 t + \varphi_0)} \sum_{q=-N}^{N} \mathrm{e}^{\mathrm{i}q(\Omega t + \eta)} = A(t)\mathrm{e}^{\mathrm{i}(\omega_0 t + \varphi_0)}, \tag{6.1.6}$$

where $A(t) = E_0 \dfrac{\sin \frac{1}{2}(2N + 1)(\Omega t + \eta)}{\sin \frac{1}{2}(\Omega t + \eta)}$ represents the time-varying electric field amplitude.

Thus, after mode-locking, the maximum light intensity $I_{\max} \propto (2N + 1)^2 E_0^2$, with a period $T_0 = \dfrac{2\pi}{\Omega} = \dfrac{2L'}{c}$. The mode-locked pulse width τ can be approximated as the time interval between the pulse peak and the first zero-intensity valley, $\tau = \dfrac{2\pi}{(2N + 1)\Omega} = \dfrac{1}{2N + 1} \dfrac{1}{\Delta\nu_q}$, the mode-locked pulse width is less than the minimum pulse width $\frac{1}{\Delta\nu_q}$ achievable with Q-switching, being only $\frac{1}{2N+1}$ of it.

To achieve mode-locking, various techniques have been developed over the past half-century, broadly categorized into two types: active mode-locking and passive mode-locking. For solid-state lasers, active mode-locking involves introducing an external periodic loss modulation mechanism, while passive mode-locking primarily utilizes the saturable absorption characteristics of materials. Representative techniques for active mode-locking include electro-optic/acousto-optic modulation mode-locking and synchronously pumped mode-locking, whereas for passive mode-locking, the main techniques are Kerr-lens mode-locking and semiconductor saturable absorber mirror (SESAM) mode-locking. For Ti:sapphire lasers, the Kerr-lens mode-locking mechanism is predominantly employed.

Kerr-lens mode-locking (KLM) is a pulse selection mechanism based on the optical Kerr effect, which is a nonlinear effect where the refractive index of a medium changes in proportion to the light intensity. Specifically, the refractive index of the medium can be expressed as

$$n = n_0 + n_2 I, \tag{6.1.7}$$

where n_0 is the refractive index independent of light intensity, n_2 is the nonlinear refractive index coefficient related to the third-order nonlinear susceptibility, and I is the pulse light intensity.

For gain media with $n_2 > 0$ (such as Ti^{3+}: Sapphire), laser pulses will undergo self-focusing due to the optical Kerr effect: high-intensity regions focus into smaller spots, while low-intensity regions focus into larger spots. By placing a small aperture in the cavity, high-intensity parts can pass through the aperture, while low-intensity parts are partially blocked and attenuated, as shown in Figure 6.4. Therefore, as the laser pulse travels

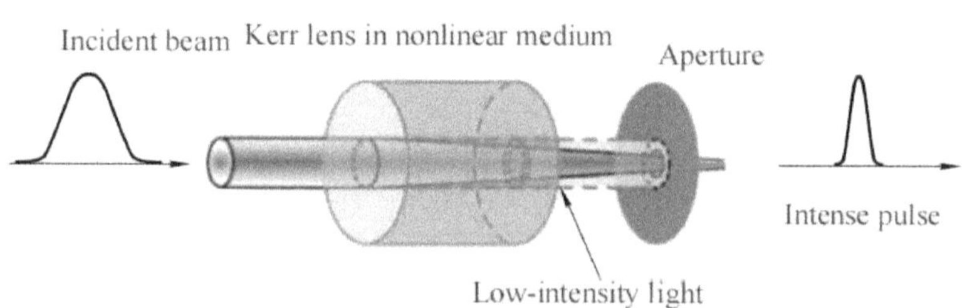

Fig. 6.4 Schematic diagram of Kerr-lens mode-locking mechanism [12].

back and forth in the cavity, low-intensity parts are continuously lost, while high-intensity parts are amplified, resulting in a narrower pulse in the time domain and ultimately producing an ultrashort mode-locked pulse. The combination of the gain medium and the cavity aperture acts as a fast saturable absorber, compressing the front and back edges of the laser pulse. In practical applications, there is often no physical small aperture; instead, the Ti:sapphire crystal serves as both the gain medium and the Kerr medium, and also functions as a soft-edge aperture, as shown in Figure 6.5.

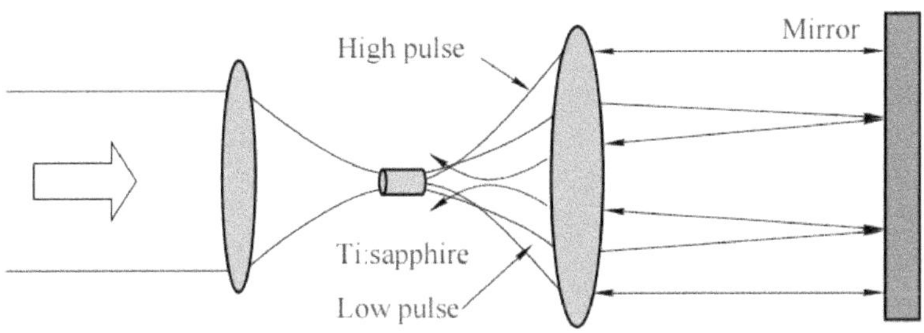

Fig. 6.5 Schematic diagram of soft-aperture Kerr-lens mode-locking.

It should be noted that the optical Kerr effect is a third-order nonlinear effect that requires a certain light intensity to manifest. Therefore, Kerr-lens mode-locking generally cannot self-start; that is, an initial perturbation must be introduced manually to initiate mode-locking. However, once mode-locking begins, the Kerr effect will sustain it continuously, and the starting mechanism is no longer needed.

6.1.3 *Dispersion Compensation*

When an electromagnetic wave propagates through a medium, the refractive index n varies with frequency ω, i.e., $n = n(\omega)$, leading to different propagation speeds and consequently a frequency-dependent phase shift $\phi(\omega)$. This phenomenon is known as dispersion. According to Fourier transform theory, a femtosecond pulse corresponds to a broad spectrum, and dispersion in a medium will cause different spectral components to move relative to each other, resulting in pulse broadening. Therefore, controlling the dispersion in a laser system is essential for generating femtosecond pulses.

To describe the dispersion effect more precisely, the phase $\phi(\omega)$ can be expanded as a Taylor series around the central angular frequency ω_0:

$$\phi(\omega) = \phi(\omega_0) + \phi'(\omega)\big|_{\omega_0}(\omega - \omega_0) + \frac{1}{2!}\phi''(\omega)\bigg|_{\omega_0}(\omega - \omega_0)^2 + \frac{1}{3!}\phi'''(\omega)\bigg|_{\omega_0}(\omega - \omega_0)^3 + \cdots, \tag{6.1.8}$$

where each term corresponds to a type of dispersion: the first-order coefficient $\phi'(\omega)$ is called group delay (GD), the second-order coefficient $\phi''(\omega)$ is called group delay dispersion (GDD), and the third-order coefficient $\phi'''(\omega)$ is called third-order dispersion (TOD), etc. Theoretically, GD is related to the temporal shift of the pulse, GDD (whether positive or negative) leads to pulse broadening, TOD and higher-order terms affect the pulse contrast, and the shortest pulse duration is achieved when the Taylor expansion of $\phi(\omega)$ has zero second-order and higher-order terms, i.e., when the pulse width reaches the Fourier-transform limit.

Dispersion compensation involves introducing dispersion of the opposite sign to the original dispersion to eliminate higher-order dispersion terms in the system and achieve ultrashort pulse output. The phase change of a light pulse passing through a medium is given by

$$\phi(\omega) = \frac{n(\omega)\omega}{c}L. \tag{6.1.9}$$

Ordinary medium materials (including Ti-doped sapphire crystals) exhibit normal dispersion, and the Taylor expansion of $n(\omega)$ around the central frequency yields positive GDD, TOD, and higher-order dispersion terms. To compensate for dispersion and compress the pulse, it is necessary to rearrange different spectral components within the laser cavity to introduce a frequency-dependent phase shift in the opposite direction (i.e., negative dispersion of various orders). Common negative dispersion optical elements in Ti:sapphire oscillators include prism pairs and chirped mirrors, which are typically used in combination in practical femtosecond oscillators.

Figure 6.6 illustrates the dispersion compensation using a prism pair. This method employs geometric effects and material dispersion for dispersion compensation. AC and BE represent the laser wavefronts. Therefore, the optical path length CDE is equal to the path length AB, i.e., the optical path length CDE is $L\cos\beta$. As shown in the figure, the angles of the various spectral components after dispersion by the prisms are slightly different, meaning that the corresponding β values for different frequencies are slightly different (although β is small and the changes in β for different frequencies are also small, the resulting optical path differences cannot

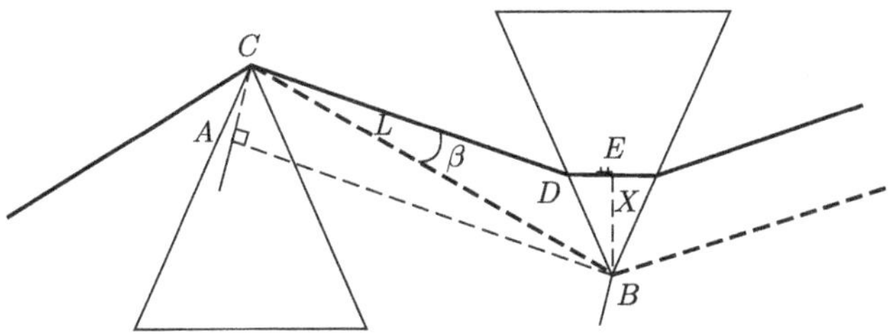

Fig. 6.6 Schematic diagram of dispersion compensation by prism.

be ignored). Consequently, the transmission paths of different wavelength spectral components between the prism pair are slightly different. According to theoretical calculations [13], by adjusting the spacing L between the two prisms and the insertion amount X of the prisms, the geometric optical path and the path length of light in the prisms can be changed, thereby quantitatively altering the sign and magnitude of the dispersion of each order of the prism pair. This allows for convenient adjustment to obtain the required dispersion compensation in practice. The main advantage of the prism pair is the continuous adjustability of the dispersion amount. The main disadvantages are that the effective compensation spectral bandwidth is not flat enough, the higher-order terms of material dispersion are difficult to eliminate, and the space occupied is relatively large, which is not conducive to miniaturization.

In addition to using geometric effects for dispersion compensation, dispersion compensation can also be achieved by designing coating layers, namely chirped mirror dispersion compensation. As shown in Figure 6.7, the chirped mirror is an extension of a multi-layer reflective mirror. The design of its coating layers is an alternating arrangement of high and low

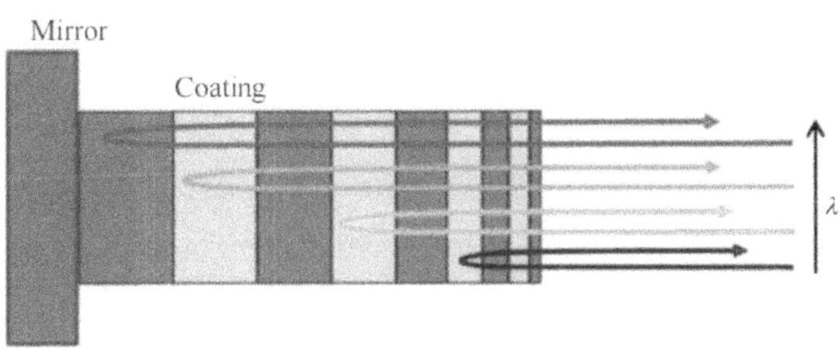

Fig. 6.7 Schematic diagram of chirped mirrors [14].

refractive index materials, but the thickness of the coating layers is not uniformly distributed and generally needs to be specifically designed to ensure that the dispersion is compensated without causing oscillations in the dispersion curve. Chirped mirrors can control the penetration depth of different wavelengths in the coating layers and can precisely design and control higher-order dispersion. They also have the advantages of low loss, compact structure, and adjustability.

6.2 Chirped Pulse Amplification (CPA) Technology

With the continuous increase in the power of Ti:sapphire lasers, nonlinear effects can occur within the Ti:sapphire crystal, leading to crystal damage. This phenomenon has limited the further increase in laser power. The introduction of chirped pulse amplification (CPA) technology has effectively solved this problem. The basic principle of CPA is as follows: For seed light generated by the oscillator with a duration of less than 10 fs and an energy level of nJ, after being broadened by a dispersive element in the time domain, the pulse is stretched to a duration of ps or even ns level. The peak power is significantly reduced, minimizing nonlinear effects during the amplification process and avoiding damage to optical components during amplification and propagation. After selection and amplification, the pulse is compressed using a compression device with a dispersion symbol opposite to that of the broadening and amplification process and a dispersion magnitude that is equivalent, resulting in a compressed pulse close to the Fourier transform limit, as shown in Figure 6.8. The CPA system mainly consists of three units: the stretcher, the amplifier, and the compressor.

6.2.1 *Stretcher*

The role of the stretcher is to broaden high-power femtosecond laser pulses in the time domain, increasing the pulse width from the fs level to the hundred ps or even ns level. This process significantly reduces the peak power density of the pulse, thereby preventing damage to components such as crystals and lenses. Currently, the most commonly used high-power stretcher is based on the geometric effect of grating diffraction to obtain positive dispersion, thereby broadening the pulse. Its initial configuration was proposed by Argentinian scholar Martinez [16] in 1987 at Bell Labs in the United States, as shown in Figure 6.9. According to theoretical analysis, when $L < f$, the structure can provide positive group delay dispersion,

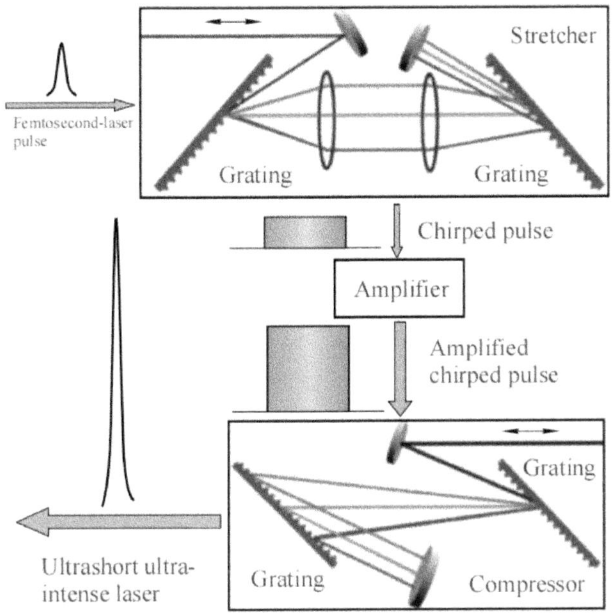

Fig. 6.8 Schematic diagram of chirped pulse amplification (CPA) [15].

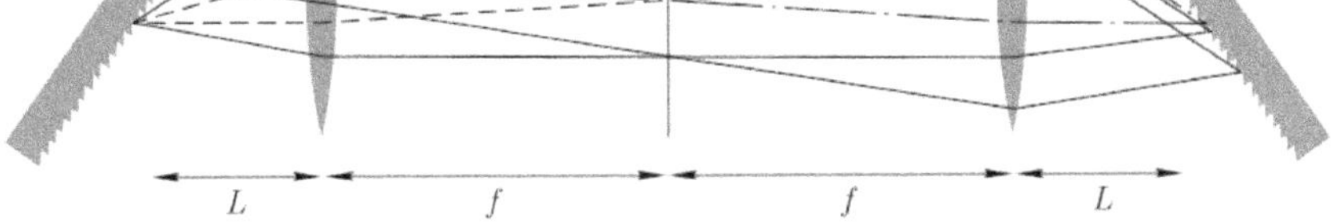

Fig. 6.9 Schematic diagram of a classic Martinez stretcher [17].

and the amount of dispersion can be adjusted. However, the presence of double lenses can introduce additional dispersion as well as aberrations and chromatic aberrations that negatively affect compression. Moreover, the double grating structure is not conducive to collimation adjustment, occupies a large space, and is costly. Therefore, an improved Martinez

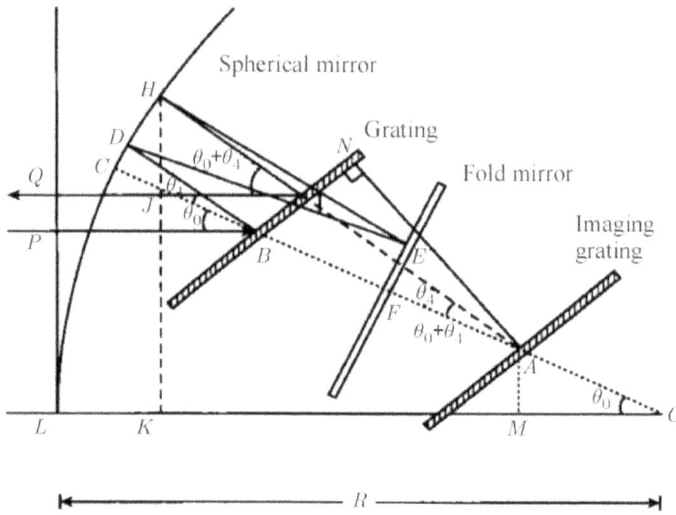

Fig. 6.10 Schematic diagram of the improved Martinez stretcher [18].

structure using a folded reflective telescope system has emerged, as shown in Figure 6.10.

The following is an analysis of how the Martinez stretcher generates positive group delayed dispersion using the ray tracing method [18]. Figure 6.10, although a folding system, is actually equivalent to the presence of two spherical mirrors. Let the distance between the two spherical mirrors be R and the axis be OC. Let the angle between the incident light and the axis OC be θ, then the diffraction angle of the grating is $\gamma - \theta = \gamma - (\theta_0 + \theta_1)$, and γ is the angle between PB and the normal of the grating. After the light is reflected twice by the spherical mirror, the incidence angle of the light to hit the grating again is

$$\gamma - (\theta_0 + \theta_4). \tag{6.2.1}$$

For the convenience of analysis, the folding system Figure 6.10 can be expanded, as shown in Figure 6.11.

By theoretical calculation [13, 18, 19], as shown in Figure 6.11, the path length of light from P to Q is

$$P = C + A - D. \tag{6.2.2}$$

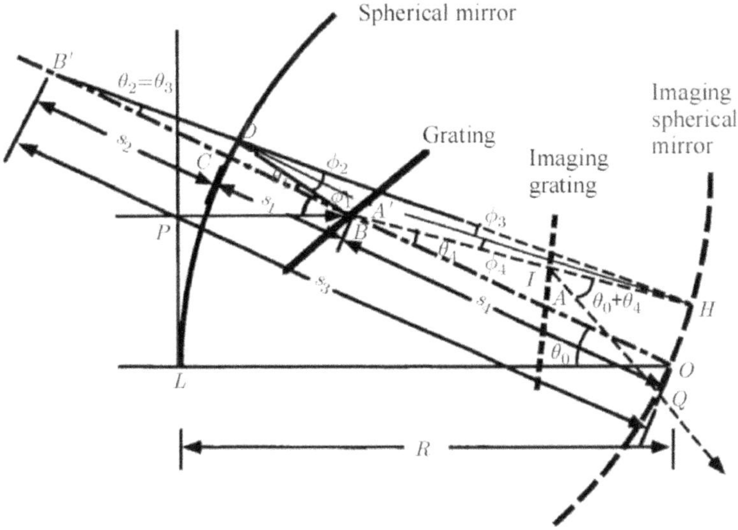

Fig. 6.11 Structural equivalence analysis diagram of the improved Martinez stretcher [19].

Among them

$$C = 2R - (R - s_1)\cos\theta_0, \tag{6.2.3}$$

$$A = R\sin(\theta_1 - \phi_1)\left(\frac{1}{\sin\theta_1} + \frac{1}{\sin\theta_2}\right) + \sin(\theta_3 - \phi_4)\cdot\left(\frac{1}{\sin\theta_3} + \frac{1}{\sin\theta_4}\right) - R\frac{\sin\phi_4}{\sin\theta_4}\cos\theta_0, \tag{6.2.4}$$

$$D = (s_4 - s_c)\frac{\cos(\gamma - \theta_0)}{\cos(\gamma - \theta_0 - \theta_4)}\cdot[1 + \cos(\theta_0 + \theta_4)] = b\cdot[1 + \cos(\theta_0 + \theta_4)]. \tag{6.2.5}$$

The section s_c is the part between the imaging grating and the spherical mirror in s_4,

$$b = (s_4 - s_c)\frac{\cos(\gamma - \theta_0)}{\cos(\gamma - \theta_0 - \theta_4)}, \tag{6.2.6}$$

where b represents the optical path length between the first grating and the second grating (the imaging grating). The total phase shift of this system is

$$\phi(\omega) = \frac{\omega}{c}(C + A - D) + \frac{2\pi G}{d}\tan(\gamma - \theta_0 - \theta_4) + \frac{2\pi}{d}(G_0 - G)\tan(\gamma - \theta_0). \tag{6.2.7}$$

The first term is the geometric phase shift, the second term is the additional phase shift due to the diffraction effect when the output beam is collimated, and the last term is the phase correction factor considering the aberration of the imaging grating. Here, G denotes the grating separation, and G_0 is the grating separation when $\theta_4 = 0$.

By performing a Taylor series expansion of Eq. (6.2.7) around the central frequency, it can be seen that the group delay dispersion term is positive, which means the pulse can be broadened. Additionally, it should be noted that the third-order dispersion term is negative.

Since the focal plane of the spherical mirror is not a plane, the Martinez-type stretcher composed of a spherical mirror and a plane folding mirror has aberrations. These aberrations eventually convert into additional dispersion, meaning it cannot be strictly conjugate with the negative dispersion compression system, and thus it is difficult to compress the pulse width back to femtoseconds. It has been analyzed that if the plane folding mirror in Figure 6.10 is replaced with a spherical mirror with a radius of $R/2$ and concentric with the original spherical mirror, an aberration-free positive dispersion element can be formed. This structure is called the Offner-type aberration-free stretcher [20]. For details on its principles and related calculations, interested readers are referred to [21], which will not be elaborated here.

6.2.2 *Amplifier*

The amplifier is the unit that amplifies the broadened laser pulse and is the core component for obtaining ultra-intense laser. Figure 6.12 shows the relationship between gain and energy extraction efficiency with the incident light (seed light) flux: the higher the energy of the incident light, the lower the gain factor, and the higher the energy extraction rate. Therefore, the best amplification effect should be achieved in stages, i.e., when the incident pulse has low energy (in the nJ range), a high-gain amplifier should be used; when the pulse energy has reached a higher level (in the mJ range), an amplifier with high energy extraction efficiency should be used. The former is commonly referred to as a preamplifier, and the latter as a power amplifier. Depending on the optical path structure, they can be further divided into regenerative amplifiers and multipass amplifiers.

A regenerative amplifier is essentially a Q-switched laser. It consists of a gain medium and a resonant cavity containing a Q-switch element, known as the regenerative cavity. The broadened seed pulse is introduced into the

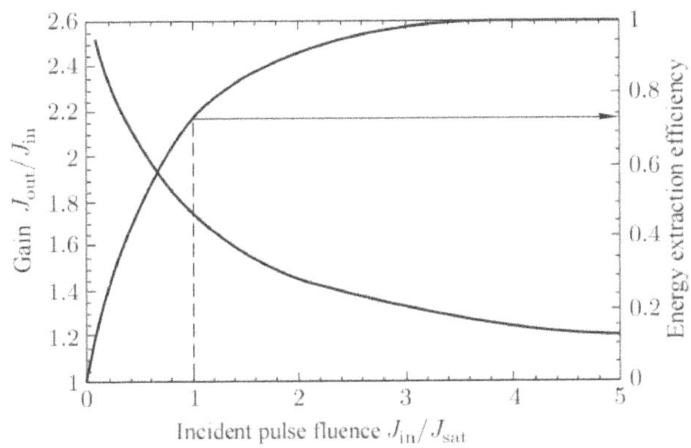

Fig. 6.12 Relation between gain and energy extraction efficiency and incident luminous flux [13].

regenerative cavity, where it passes through the crystal multiple times and is amplified until saturation. Then, by controlling the Q-switch element, the pulse is directed out of the cavity. Depending on the type of voltage required for the Q-switch element, it can be divided into quarter-wave voltage type and half-wave voltage type. The former has lower performance requirements for the Q-switch device: the required voltage does not have to be too high, and the rise time does not have to be too short; the latter can completely isolate the incident light and the output light, avoiding the influence of back-reflected light on the previous stretcher and oscillator. Depending on the cavity structure, it can also be divided into linear regenerative cavities and ring regenerative cavities. The former is a standing-wave amplifier, with the advantages of simple structure, easy adjustment, and short pulse establishment time; the latter is a traveling-wave amplifier, with the main advantage of effectively reducing amplified spontaneous emission (ASE) and improving pulse contrast.

The most common Q-switch element is a Pockels cell, which is an electro-optic device with an electro-optic crystal at its core (such as KDP, DKDP, ADP, KDA, etc.). By applying voltage to the electro-optic crystal, the Pockels cell can provide fast and precise control of the output light polarization direction that changes with the applied voltage. The working principle of the Pockels cell is based on the Kerr electro-optic effect: applying a constant or varying voltage (electric field) to the electro-optic crystal can cause a linear change in the crystal's birefringence, thereby changing the polarization state of the passing light beam.

Here, taking a linear quarter-wave voltage-type regenerative amplifier as an example, the basic working principle of the regenerative amplifier is briefly described.

Figure 6.13 shows a schematic diagram of a linear quarter-wave voltage-type regenerative amplifier structure, where TFP is a thin-film polarizer (reflecting vertically polarized light and transmitting horizontally polarized light), $\lambda/4$ is a quarter-wave plate (this plate can be omitted if the Pockels cell is adjusted to a quarter-wave plate state in advance), and PC is the Pockels cell.

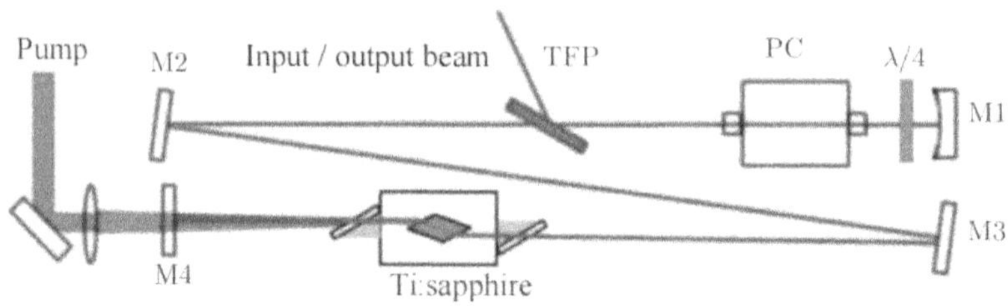

Fig. 6.13 Structure diagram of a linear quarter-wave voltage regenerative amplifier [22].

The vertically polarized seed light is reflected by the TFP into the regenerative cavity. At this time, the PC is not powered, so the seed light passes through the PC, the quarter-wave plate, is reflected by M1, and then passes through the quarter-wave plate again to become horizontally polarized light. It can then pass through the PC and the TFP, and after being reflected by M2, M3, the gain crystal, and M4, it returns along the original path. Until it passes through the quarter-wave plate twice again to become vertically polarized, it then passes through the PC and is reflected by the TFP to exit the cavity. In this entire process, the pulse can only be amplified twice, which is far from saturation. If a quarter-wave voltage is applied to the PC after the seed light becomes horizontally polarized and before it returns to the PC for the next time, the seed light will pass through the PC and the quarter-wave plate, which is equivalent to passing through a half-wave plate, and thus it becomes vertically polarized. It is then reflected by M1 and passes through the two components again to become horizontally polarized. Therefore, it can continue to oscillate and be amplified within the cavity. Until the pulse energy in the cavity reaches its maximum (usually requiring several to dozens of round trips), to extract the pulse, simply remove the voltage applied to the PC. The pulse will then pass through the PC and the quarter-wave plate again, be reflected by M1, and return to the TFP, becoming vertically polarized, and can then be directed out of the cavity by the TFP. Thus, the function of amplifying the energy of the broadened seed pulse is completed.

A multipass amplifier utilizes mirrors to reflect the incident light multiple times through the gain crystal and extract energy from it. Clearly, the multipass amplifier uses fewer transmissive elements, thus introducing less material dispersion and resulting in shorter amplified pulses. The most typical application of the multipass amplifier is in the amplification of low-repetition-rate (< 100 Hz) and high-energy (> 1 mJ) laser pulses, where its structure is relatively fixed, i.e., the bow-tie optical path, as shown in Figure 6.14.

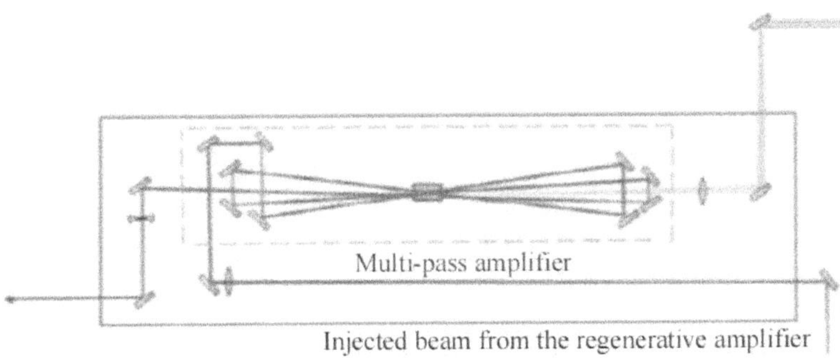

Fig. 6.14 Bow type optical path of multi-pass amplifier [22].

It should be noted that a multipass amplifier can also be used in the preamplification stage. Whether a multipass amplifier or a regenerative amplifier is superior depends on the requirements of the amplified and compressed pulse [13]. Generally, in cases where the pulse width is not too narrow (above 20 fs), the regenerative amplifier has a significant advantage. When the pulse width is very narrow (< 20 fs), the multipass amplifier is more commonly used. In situations where very high pulse energy is required, both types are often used in combination, i.e., the preamplifier uses a regenerative amplifier, and the power amplifier uses a multipass amplifier. Considering the high stability of high-repetition-rate pump lasers, a high-repetition-rate regenerative amplifier can be used in the preamplification stage, and a low-repetition-rate multipass amplifier can be used in the power amplification stage, with a Pockels cell pulse selection device added in between. The specific parameter comparison between the two is shown in Table 6.2.

Table 6.2 Comparison of regenerative amplifiers and multipass amplifiers [13].

Parameter	Regenerative Amplifier	Multi-Pass Amplifier
Beam quality	The output beam can be considered Gaussian since the regenerative amplifier itself operates in TEM00 mode, independent of the input pulse's beam quality.	The multi-pass amplifier replicates the input pulse's beam, and non-collinear incidence introduces additional non-uniformity.
Pointing stability	The pointing stability depends solely on the amplifier cavity's stability. Oscillator and stretcher pointing variations do not affect the output beam.	Pointing stability inevitably traces back to the oscillator's stability. Pump energy fluctuations also cause slight thermal lens changes in the gain medium, leading to significant spatial beam variations.
Energy stability	High stability is easily achieved by increasing the number of round trips to ensure gain saturation.	Pump energy must be increased to ensure gain saturation, but adding more passes is challenging.
Reliability	The number of round trips can be adjusted almost indefinitely, and the crystal does not require extremely high pump energy, reducing the risk of damage.	To achieve very high gain (e.g., amplifying pulses from nJ to mJ in 8 passes), extremely high pump energy is needed, which easily damages the crystal.
Adaptability	The stretcher can be designed to accommodate significant material dispersion. Adding external optics only requires minor compressor adjustments.	The stretcher is designed for minimal material dispersion, so adding optics causes large pulse width variations.
Pulse compressibility	The stretcher and compressor can be designed to accommodate the amplifier's material dispersion, enabling compression to near-transform-limited levels.	Due to smaller material dispersion, the stretcher can support broader bandwidths, allowing shorter pulse compression.
Energy efficiency	Efficiency depends on the cavity's performance, typically up to 20%.	Non-collinear amplification results in lower efficiency compared to regenerative amplifiers.
Complexity	Generally simpler design and requires less alignment time.	More optical components, larger space requirements, and longer alignment times are needed.

6.2.3 *Compressor*

To compress the chirped pulse back to a pulse width comparable to the seed pulse after amplification, a negative dispersion system with a dispersion amount equivalent to that of the stretcher but with an opposite sign is needed. A widely used structure is the grating pair structure proposed by Treacy in 1969 [23], as shown in Figure 6.15.

To calculate the dispersion of this system [13], first determine the path length of the light in the system: assuming the incident angle is γ and the diffraction angle is $\gamma - \theta$, the relationship between them follows the grating equation,

$$\sin \gamma + \sin(\gamma - \theta) = m\lambda/d, \tag{6.2.8}$$

where m is the diffraction order and d is the grating constant. In the grating pair structure shown in Figure 6.15, the path length P of the light ray ABC

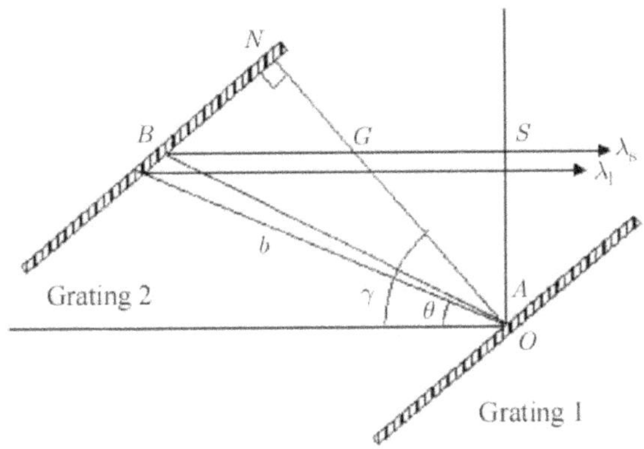

Fig. 6.15 Treacy structural grating pair compressor [13].

can be written as

$$P = b(1 + \cos\theta) = \frac{G}{\cos(\gamma - \theta)}(1 + \cos\theta). \tag{6.2.9}$$

The actual phase is not only $\phi = \omega P/c$ but also includes a phase correction factor. This is because the diffracted light from the first grating is not simply reflected when collimated by the second grating; instead, it is diffracted. Different wavelength components have not only path length differences but also phase differences due to different diffraction positions. In Figure 6.15, if the foot of the perpendicular N is taken as the reference point, then the phase correction factor for any wavelength component can be written as the number of grooves between NB multiplied by 2π, i.e., $2\pi\frac{G\tan(\gamma - \theta)}{d}$. Therefore, the total phase is

$$\phi(\omega) = \frac{\omega}{c}b(1 + \cos\theta) - \frac{2\pi G}{d}\tan(\gamma - \theta). \tag{6.2.10}$$

By performing a Taylor series expansion of Eq. (6.2.10) around the central frequency, it can be seen that the group delay dispersion is negative, while the third-order dispersion is positive, which is complementary in sign to the stretcher results in Section 6.2.1. Therefore, this grating pair can be used to compress the broadened and amplified pulse.

The above sections have provided a brief introduction to the basic principles of a typical Ti:sapphire femtosecond laser system. In Section 6.3, we will review the status of typical PW laser systems worldwide.

6.3 Typical PW Laser Systems Worldwide

With the widespread application of high-peak-power lasers, numerous hundred-TW laser systems have been established worldwide, and nearly 50 PW (10^{15} W) laser systems have been completed or are under construction [24, 25]. The distribution of laser facilities with powers ranging from 10 to 100 PW (both existing and planned) is shown in Figure 6.16. The following is a brief introduction to some typical PW laser systems worldwide.

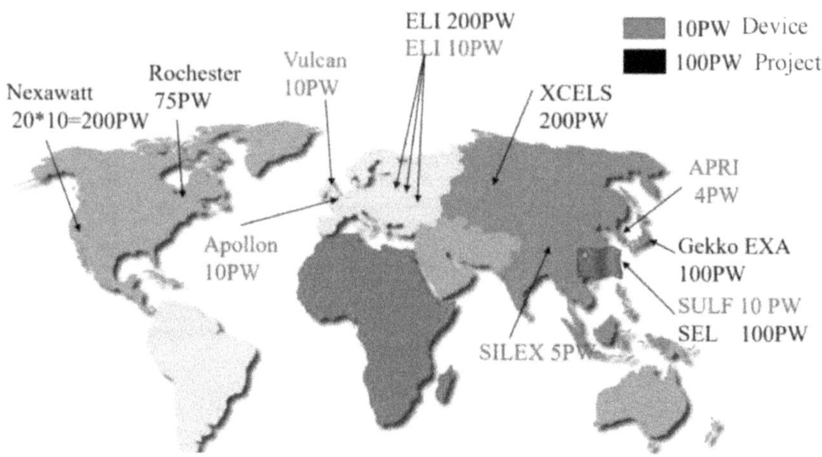

Fig. 6.16 Distribution of laser devices (including plans) from 10 to 100 PW.

6.3.1 *BELLA PW Laser System*

In 2012, the Lawrence Berkeley National Laboratory (LBNL) in the United States completed the world's first commercial PW laser system [26], as shown in Figure 6.17. The laser outputs 46 J of energy with a pulse width of 45 fs at a repetition rate of 1 Hz. This laser system, provided by Thales of France, is used for research on laser-plasma acceleration and related plasma interaction physics. Using the BELLA system, LBNL has achieved an electron beam energy of 7.8 GeV [27].

The BELLA PW laser facility is based on a dual Ti:sapphire CPA amplifier. The system composition and parameters are shown in Figure 6.18. The main amplifier is pumped by 12 Nd:YAG beams, each with an energy of 13 J. Between the two CPA amplifiers, cross-polarized wave (XPW) filtering

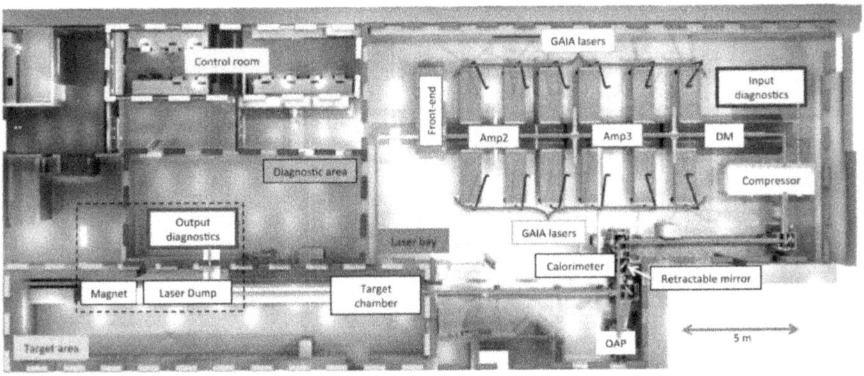

Fig. 6.17 BELLA PW laser device [26].

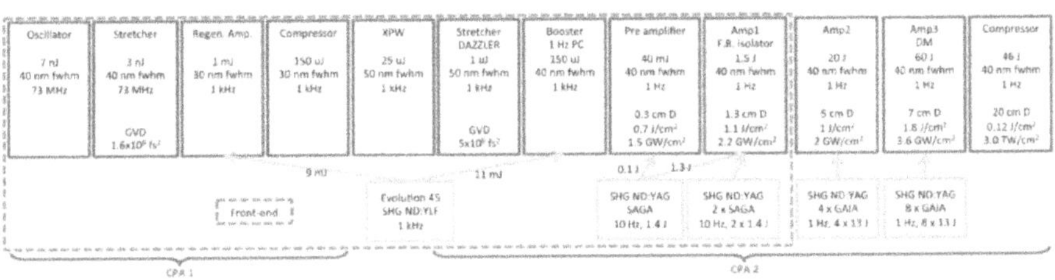

Fig. 6.18 Components and parameters of the BELLA PW laser device [26].

technology is utilized to enhance pulse contrast and broaden the spectrum. The XPW system employs two BaF_2 crystals to improve efficiency (17%). To avoid back-reflected light, the BaF_2 crystals are designed with a wedge shape. A deformable mirror for wavefront adjustment is placed between Amp3 and CPA2. The stability of the amplified pulse energy output fluctuates by 0.54%.

6.3.2 *DRACO PW Laser System*

The DRACO (Dresden Laser Accelerator) PW laser system [28], provided by Amplitude Technologies, is primarily used for plasma accelerator research. The system can produce two output beams: 150 TW and 1 PW. The basic structure of the Ti:sapphire laser system is shown in Figure 6.19. The signal from the femtosecond oscillator is pre-amplified, compressed, purified by nonlinear XPW pulse cleaning, and then amplified to a pulse energy of 1.5 J through CPA. After further amplification, the pulse energy

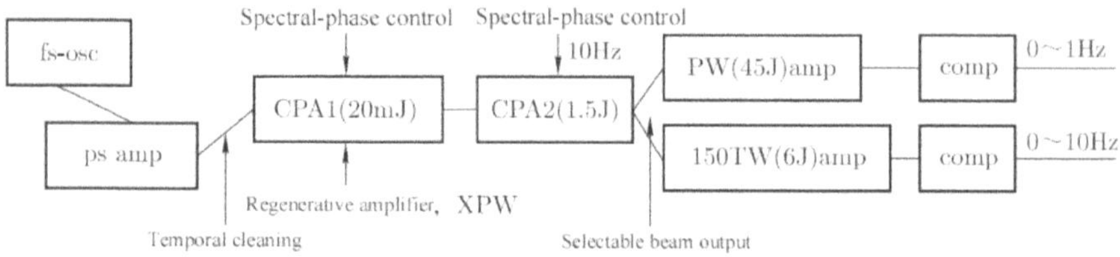

Fig. 6.19 DRACO PW laser system [28].

reaches 45 J, operating at 1 Hz with a pulse width of 30 fs. The pulse is then transmitted to various experimental areas through a wavefront correction system. The pulse time contrast is further enhanced using a plasma mirror.

6.3.3 *APOLLON PW Laser System*

The French APOLLON PW laser system [29] is a hybrid amplification system combining OPCPA and Ti:sapphire CPA, aiming to achieve high-energy short pulses (10 PW, 15 fs, 150 J, single shot per minute) with high contrast output. The design of APOLLON includes four independent laser outputs, as shown in Figure 6.20.

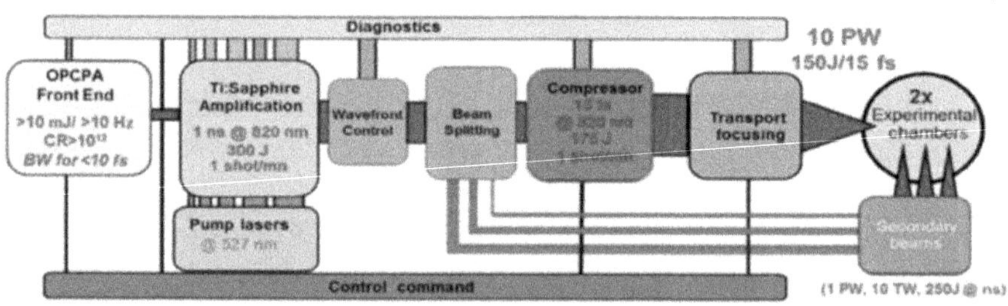

Fig. 6.20 APOLLON PW laser device [29].

6.3.4 *ELI Facility*

The Extreme Light Infrastructure (ELI) [30], a future large-scale scientific facility in the European Union, comprises four major research installations: the ELI-Beamlines Facility in Prague, Czech Republic; the ELI-Attosecond Facility in Szeged, Hungary; the ELI-Nuclear Physics Facility in Magurele, Romania; and the ELI-Ultra High Field Facility. ELI will provide a unique platform for users of ultra-high-power lasers, with multiple laser system facilities configured in dedicated target areas.

(1) ELI Beamline Facility

The ELI Beamline Facility will offer a range of laser systems for research purposes in physics and materials science, biomedicine, and laboratory astrophysics. The laser systems are based on OPCPA technology, Ti:sapphire CPA technology, or a combination of both. The facility consists of four laser systems, as shown in Figure 6.21, with the following output energies and repetition rates:

L1: 100 mJ / 1 kHz

L2: 1 PW / 20 J / 10 Hz (laser radiography beamline)

L3: 1 PW / 30 J / 10 Hz (laser system HAPLS built by the Lawrence Livermore National Laboratory (LLNL), using diode-pumped solid-state laser (DPSSL) pumped Ti:sapphire CPA technology)

L4: 10 PW / 1.5 kJ / 1 shot per minute.

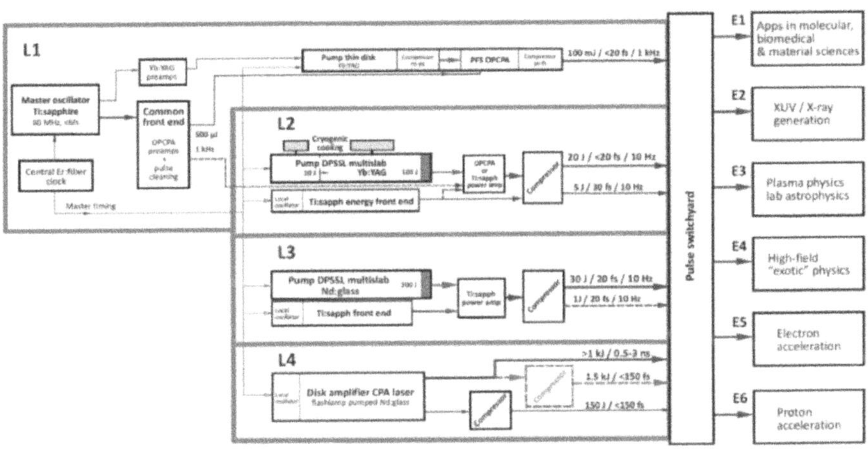

Fig. 6.21 Schematic diagram of ELI bundling device.

(2) ELI Nuclear Physics Facility

The ELI-Nuclear Physics Facility (ELI-NP) in Romania features a high-power laser system (High-Power Laser System, HPLS) with two 10 PW laser systems. The HPLS at ELI-NP combines OPCPA based on low-energy levels and large-aperture Ti:sapphire amplifiers pumped by high-power Nd:YAG and neodymium glass frequency-doubled lasers. The system design is shown in Figure 6.22. The front end consists of OPCPA,

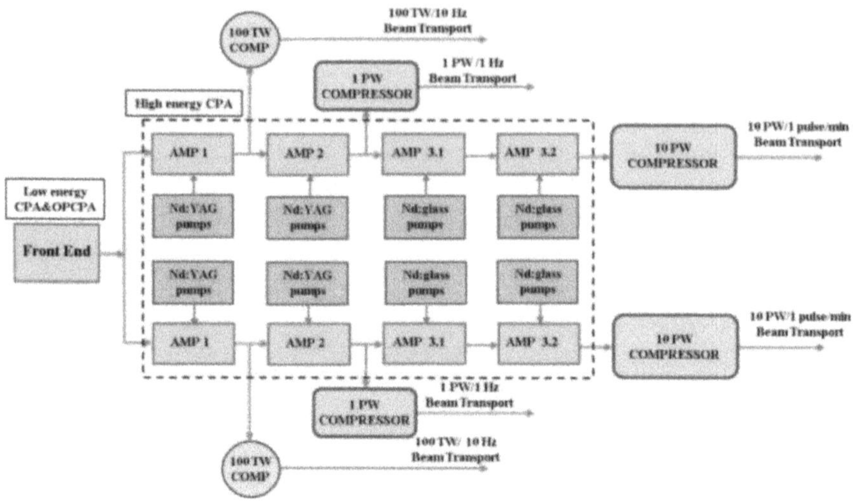

Fig. 6.22 ELI-NP dual 10 PW laser system.

while the main amplifier is composed of Ti:sapphire amplifiers. The laser system outputs: 2×100 TW, 10 Hz, 2×1 PW, 1 Hz, 20 J, < 20 fs, 2×10 PW, single shot per minute, 250 J, 25 fs. The beam diameter is 50 cm (achieving a focused power density of 10^{23} W/cm^2), used for high-energy physics experiments under extreme conditions, light-induced nuclear physics reactions and their applications (including nuclear fission and fusion), and accelerating electrons to over 10 GeV.

(3) ELI Attosecond Light Pulse Source (ALPS)

The ELI-ALPS will provide three high-repetition-rate OPCPA light sources and strong-field laser sources, namely:

High-repetition-rate laser: 100 kHz, > 5 mJ, < 6 fs

Mid-infrared laser: 10 kHz, > 10 mJ, $4 \sim 8$ μm

Terahertz-pumped laser: 100 Hz, > 1 J, < 5 fs

Single-cycle (SYLOS) laser: 1 kHz, > 100 mJ, < 5 fs

Strong-field laser: 10 Hz, > 2 PW, < 10 fs.

All these sources will be used to drive auxiliary light sources (UV/XUV, X-rays, ions, etc.) and will be dedicated to the study of electron dynamics in atoms, molecules, plasmas, and solids.

6.3.5 *CoReLS PW Laser System*

The CoReLS PW laser system [31] at the Korea Advanced Institute of Science and Technology uses a hybrid OPCPA and CPA amplification technique. The compressed pulse width is 19.4 fs, the output pulse energy is 83 J, the energy stability is 1.5% (rms), and the peak power reaches 4.2 PW. The laser system composition is shown in Figure 6.23. The system introduces XPW and OPCPA technologies to prevent spectral gain narrowing during amplification and enhances the temporal contrast. Thanks to the XPW technique, the temporal contrast at the 100 ps level reaches 3×10^{-12}, which is four orders of magnitude higher than the front end's 2×10^{-8}.

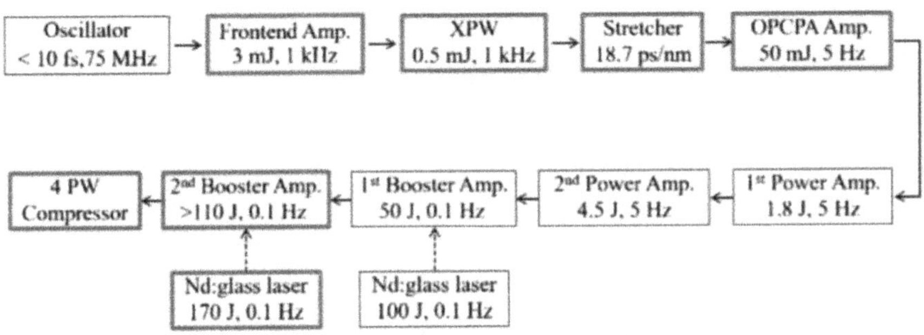

Fig. 6.23 CoReLS 0.1 Hz, 4.2 PW CPA titanium gem laser system [31].

6.3.6 *SULF PW Laser System*

The SULF PW laser system [32, 33] at the Shanghai Institute of Optics and Fine Mechanics is a dual CPA amplification chain system based on Ti:sapphire crystals with a central wavelength of 800 nm. The system uses XPW and optical parametric amplification (OPA) modules to achieve wide-spectrum, high-contrast laser pulse output.

The laser system principle is shown in Figure 6.24. The output laser parameters are: pulse width of 24 fs, spectral bandwidth (FWHM) greater than 60 nm, compressor efficiency of 64%, compressed pulse energy of 216.96 J, corresponding to a laser pulse peak power of 10.3 PW.

6.3.7 *Peking University PW Laser System*

The Peking University Laser Plasma Laboratory, in collaboration with Thales of France, developed a 2 PW laser system based on OPCPA and

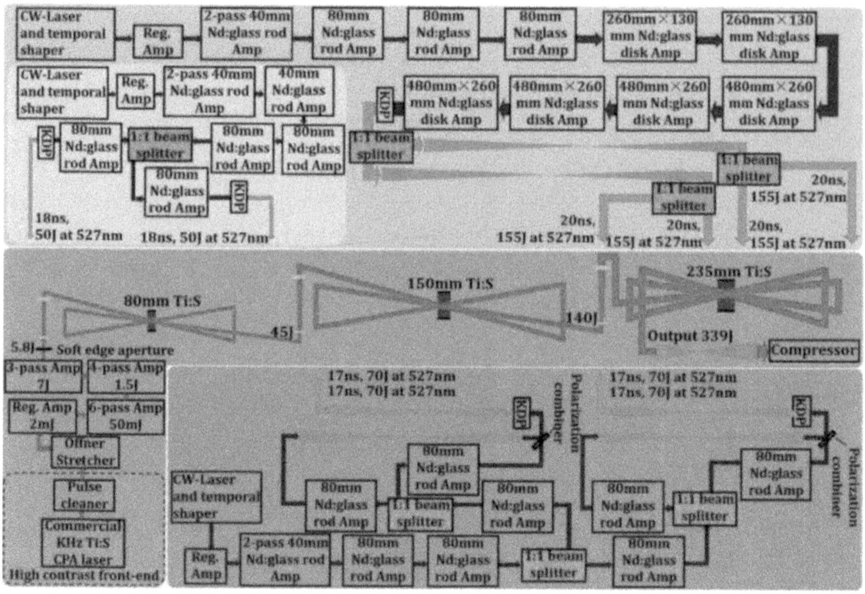

Fig. 6.24 Schematic diagram of SULF [32].

XPW technologies. The peak power is 2 PW, the laser output energy is ≥ 60 J, the pulse width is ≤ 30 fs, and the repetition rate is 1 Hz. This laser system is primarily used for the development of laser proton accelerators.

References

[1] Maiman T. H. Stimulated optical radiation in ruby [J]. *Nature*, 1960, 187(4736): 493–494.

[2] Collins R. J., Kisliuk P. Control of population inversion in pulsed optical masers by feedback modulation [J]. *Journal of Applied Physics*, 2004, 33(6): 2009–2011.

[3] Hargrove L. E., Fork R., Pollack M. A. Locking of He-Ne laser modes induced by synchronous intracavity modulation [J]. *Applied Physics Letters*, 1964, 5: 4–5.

[4] Fork R. L., Greene B. I., Shank C. V. Generation of optical pulses shorter than 0.1 psec by colliding pulse mode locking [J]. *Applied Physics Letters*, 1981, 38(9): 671–672.

[5] Moulton P. Ti-doped sapphire: tunable solid-state laser [J]. *Optics News*, 1982, 8(6): 9-9.

[6] Spence D. E., Kean P. N., Sibbett W. 60-fsec pulse generation from a self-mode-locked Ti:sapphire laser [J]. *Opt. Lett.*, 1991, 16(1): 42–44.

[7] Strickland D., Mourou G. Compression of amplified chirped optical pulses [J]. *Optics Communications*, 1985, 56(3): 219–221.

[8] `https://gb.castech.com/product/Ti%3ASapphire-%E6%99%B6%E4%BD%93-85.html` [Z].

[9] Moulton P. F. Spectroscopic and laser characteristics of Ti:Al2O3 [J]. *J. Opt. Soc. Am. B*, 1986, 3(1): 125–133.

[10] Li Yan, Yang Hong. From ultra-short light to ultra-short ultra-intense light: Interpreting the chirped pulse amplification technology awarded the 2018 Nobel Prize in Physics [J]. *Physics and Engineering*, 2019, 29(2): 3–7.

[11] Zhou Bingkun, Gao Yizhi, Chen Tirong, *et al.* Laser Principles [M]. Beijing: National Defense Industry Press, 2014.

[12] Keller U. Recent developments in compact ultrafast lasers [J]. *Nature*, 2003, 424(6950): 831–838.

[13] Zhang Zhigang. Femtosecond Laser Technology [M]. Beijing: Science Press, 2011.

[14] Yefet S., Pe'er A. A review of cavity design for Kerr lens mode-locked solid-state lasers [J]. *Applied Sciences*, 2013, 3(4): 694–724.

[15] Majinggui, Wang Jing, Qian Liejia. The Foundational Invention of Femtosecond Laser Amplification and the Breakthrough of Its Capability Limits [J]. *Physics*, 2018, 47(12): 772–778.

[16] Martinez O. 3000 times grating compressor with positive group velocity dispersion: application to fiber compensation in 1.3–1.6μm region [J]. *IEEE Journal of Quantum Electronics*, 1987, 23(1): 59–64.

[17] `https://en.wikipedia.org/wiki/Chirped_pulse_amplification#/media/File`: Cpa stret cher.svg [Z].

[18] Zhang Zhigang, Sun Hong. Calculation and Evaluation Methods of Dispersion in Femtosecond Pulse Amplifiers [J]. *Acta Physica Sinica*, 2001, 50(6): 1080–1086.

[19] Zhang Z., Yagi T., Arisawa T. Ray-tracing model for stretcher dispersion calculation [J]. *Appl. Opt.*, 1997, 36(15): 3393–3399.

[20] Cheriaux G., Rousseau P., Salin F., *et al.* Aberration-free stretcher design for ultrashort pulse amplification [J]. *Opt. Lett.*, 1996, 21(6): 414–416.

[21] Jiang J., Zhang Z., Hasama T. Evaluation of chirped-pulse-amplification systems with Offner triplet telescope stretchers [J]. *J. Opt. Soc. Am. B*, 2002, 19(4): 678–683.

[22] He Peng. High Average Power Femtosecond Ti:Sapphire Lasers and the Generation and Control of Sub-Cycle Optical Pulses [D]. Xi'an: Xidian University, 2017.

[23] Treacy E. Optical pulse compression with diffraction gratings [J]. *IEEE Journal of Quantum Electronics*, 1969, 5(9): 454–458.

[24] Danson C., Hillier D., Hopps N., *et al.* Petawatt class lasers worldwide [J]. *High Power Laser Science and Engineering*, 2015, 3: e3.

[25] Danson C. N., Haefner C., Bromage J., *et al.* Petawatt and exawatt class lasers world wide [J]. *High Power Laser Science and Engineering*, 2019, 7: e54.

[26] Nakamura K., Mao H. S., Gonsalves A. J., *et al.* Diagnostics, control and performance parameters for the BELLA high repetition rate petawatt class laser [J]. *IEEE Journal of Quantum Electronics*, 2017, 53(4): 1–21.
[27] Gonsalves A. J., Nakamura K., Daniels J., *et al.* Petawatt laser guiding and electron beam acceleration to 8 GeV in a laser-heated capillary discharge waveguide [J]. *Physical Review Letters*, 2019, 122(8): 084801.
[28] Schramm U., Bussmann M., Irman A., *et al.* First results with the novel petawatt laser acceleration facility in Dresden [J]. *Journal of Physics: Conference Series*, 2017, 874(1): 012028.
[29] Papadopoulos D. N., Zou J. P., Le Blanc C., *et al.* First commissioning results of the Apollon laser on the 1 PW beam line [C]. CLEO: Science and Innovations. Optica Publishing Group, 2019: STu3E. 4.
[30] `https://www.eli-beams.eu/` [Z].
[31] Sung J. H., Lee H. W., Yoo J. Y., *et al.* 4.2 PW, 20 fs Ti:sapphire laser at 0.1 Hz [J]. *Opt. Lett.*, 2017, 42(11): 2058–2061.
[32] Li W., Gan Z., Yu L., *et al.* 339 J high-energy Ti:sapphire chirped-pulse amplifier for 10 PW laser facility [J]. *Opt. Lett.*, 2018, 43(22): 5681–5684.
[33] Gan Z., Yu L., Wang C., *et al.* The Shanghai Superintense ultrafast Laser Facility (SULF) project [J]. *Progress in Ultrafast Intense Laser Science XVI*, 2021: 199–217.

Chapter 7

Preparation of Laser Ion Acceleration Targets

Experimental studies have shown that in laser ion acceleration, the thickness, density, surface morphology, and geometric dimensions of the target material directly affect the energy, energy spectrum, and angular distribution of ions. Therefore, target preparation is a very important aspect of laser ion acceleration research. This chapter will provide an overview of the main targets and their preparation methods in the field of laser ion acceleration.

Targets for laser ion acceleration can be categorized by their form into solid thin film targets, foam targets, micro- and nano-structured targets, and liquid targets; by density into solid targets and near-critical density targets; and by composition into organic polymer (plastic) targets, metal targets, and inorganic compound targets. The following sections introduce these target types according to their form.

7.1 Solid Thin Film Targets

7.1.1 *Plastic Thin Film Targets*

Plastic thin film targets are widely used in laser ion acceleration. Compared with other types of targets, plastic thin film targets have the following advantages: (1) The thickness range of the target material is relatively large (10 nm to 10 μm), which means a wide range of applications. Considering laser contrast as well as target energy and pulse width, different experiments have different requirements for the thickness of plastic thin film targets. Due to the large spatial freedom of molecular chains and strong chain flexibility of polymers, they exhibit excellent ductility and good mechanical strength on a macroscopic scale, and can be self-supported over a wide thickness range, making them highly applicable targets. Currently, using polymers, thin film targets thinner than 10 nm can be prepared, making them ideal targets for laser ion acceleration in the Radiation Pressure Acceleration (RPA) mechanism. (2) The preparation process is simple and has low requirements for preparation equipment, allowing for the mass production of a large number of targets in a short time. As a consumable in laser acceleration experiments, plastic thin film targets are inexpensive, which helps control costs — this is of great significance for the application of laser accelerators. The main preparation methods for plastic thin film targets currently include spin coating, dip coating, drop casting, and extrusion (see Figure 7.1).

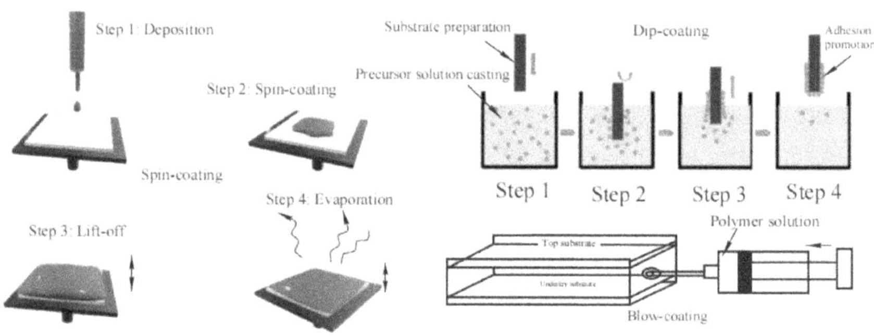

Fig. 7.1 Schematic diagram of plastic film preparation [27–29].

Spin coating is a relatively simple and common method. It utilizes the centrifugal force to spread the organic polymer solution on a substrate, resulting in a thin film with a thickness in the range of nm to μm. Spin coating can be performed on silicon wafer surfaces or other materials with affinity. Laser acceleration experiments require the thin film to be self-supported in a certain area without relying on any substrate. To achieve this, the main technical route currently adopted by spin coating is: first, a "sacrificial layer" is coated on the substrate surface (such as a glass slide or silicon wafer), then the plastic film is spin-coated on top, and the substrate with the film is slowly immersed in water. At this point, the sacrificial layer dissolves in the solvent, and the plastic film automatically floats on the liquid surface. A perforated target sheet is used to retrieve the desired film (as shown in Figure 7.2).

The most commonly used solvent for de-molding is water, and the corresponding sacrificial layer is mainly composed of two major categories of materials: one is amphiphilic polymers that are both lipophilic and

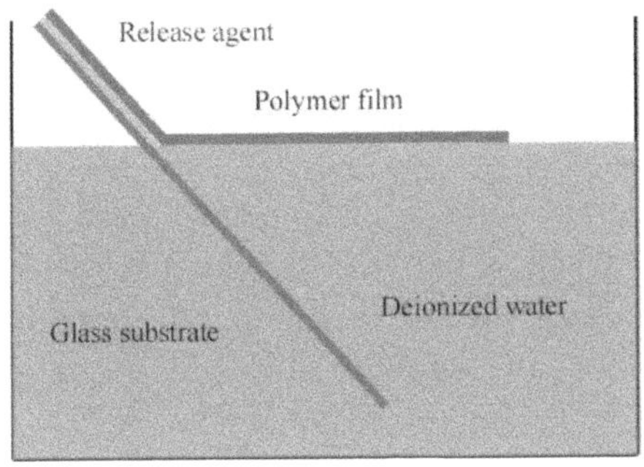

Fig. 7.2 Production of self-supporting plastic film [30].

hydrophilic (such as soapy water), whose hydrophilic groups can dissolve in water to separate the film from the substrate; the other is inorganic materials that are easily soluble in water (see Section 7.1.2 on metal film preparation methods). The film target itself is insoluble in water, while the sacrificial layer is soluble in water, thus naturally achieving the separation of the film and the substrate.

The advantage of spin coating is that it can quickly produce a large number of self-supported plastic films with good uniformity. Using low-concentration solutions, ultra-thin plastic films can also be prepared. Its disadvantage is that the area of the film obtained by spin coating is often limited, which is not conducive to the large-area preparation of targets. Moreover, when the concentration of the organic solution is high, the viscosity will also be high, and the centrifugal force will find it difficult to spread the solution evenly into a film, so this method has difficulty in preparing films thicker than μm over a large area.

Drop casting involves dropping a prepared organic solution that is insoluble in water onto the surface of water using a liquid volume-controlled pipette. The organic solution spreads and extends under the action of surface tension, and after a few seconds, the solvent evaporates to form a plastic film. The film can be retrieved using a ?? to produce a self-supported plastic film target. Using a multi-channel array pipette, large-area plastic films can be prepared on the water surface. Drop casting combines the "de-molding - floating" processes into one, shortening the intermediate steps, making the operation more convenient and quick, and can provide a large number of targets for high-repetition-rate target shooting. Currently, the solutes used for drop casting plastic film targets include polyvinyl butyral resins, polystyrene, xylene, pyridine, etc., and the solvents generally used are ethanol, 1,2-dichloroethane, acetone, chloroform, etc. Sebastian Seuferling *et al.* [2] have shown that polyvinyl alcohol formaldehyde (PVA) in 1,2-dichloroethane solution and polystyrene in chloroform solution yield good experimental results. Although the process of drop casting plastic films is simple, the thickness of the film can vary greatly due to uneven solute diffusion and internal stress, and the flatness is poor (as shown in Figure 7.3), which needs further research and improvement.

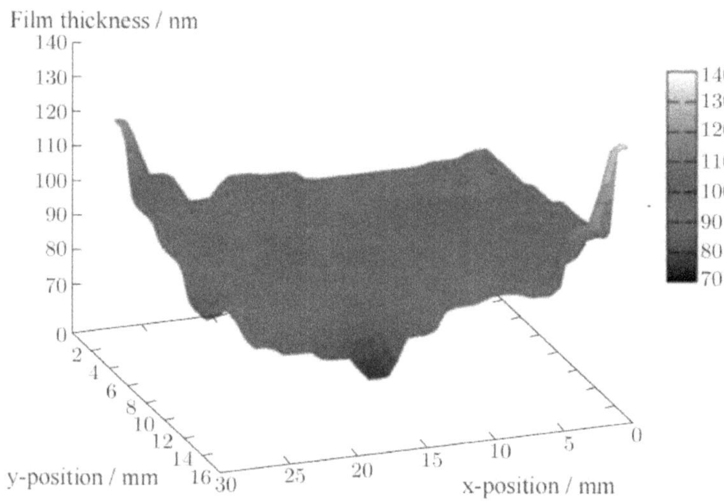

Fig. 7.3 Thickness distribution of thin films measured by spectral method [2].

Immersion and draw-up method involves inserting a modified ?? into the organic solution, and then slowly drawing it out at a uniform speed after the solution adsorbs on the surface of the ?? through contact and immersion. The film is then dried in an oven at an appropriate temperature and atmosphere to form a self-supported film.

Slit extrusion involves using an organic solution with a high surface tension to coat and adsorb on the substrate surface through blow molding or extrusion, and then drying it into a film. This process is relatively cumbersome and results in a thicker film, which is mostly used for the preparation of commercial plastic films. In laser acceleration research, if a thicker film (μm thickness) is needed, commercial films can be purchased, stretched, and directly attached to the target sheet as the target body.

Characterization of the prepared plastic films is also an important part of target preparation. For common plastic films, it is necessary to know their thickness and flatness information. For films closely attached to the substrate, optical profilometers or step meters, atomic force microscopes, etc., can be used to measure their thickness. Specifically, a sharp blade is first used to scratch a line on the film. The film at the scratched area naturally peels off. The surface morphology around the scratch is measured with an optical profilometer or other equipment, and the thickness can be obtained through the step height difference. Figure 7.4 shows the three-dimensional (3D) morphology around the scratch measured by an optical profilometer, from which the thickness value can be obtained. For self-supported plastic films, the flatness can be characterized using an optical profilometer.

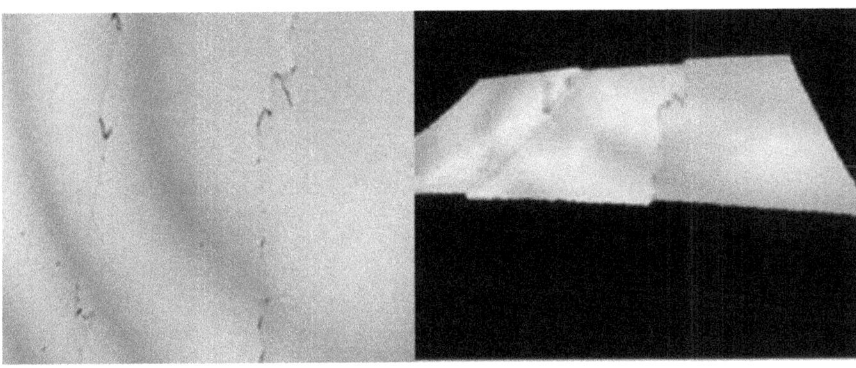

Fig. 7.4 Micrograph (left) and 3D surface rendering diagram (right) of the scratched plastic film under the optical morphometer.

7.1.2 *Metal Thin Film Targets*

Metal thin film targets are also widely used in laser acceleration experiments. Compared with plastic targets, metal targets have a larger mass number and higher plasma density, and can be used to accelerate heavy ions. Heavy ion beams have a wide range of applications in fields such as fast ignition of nuclear fusion, irradiation phase transformation, research on special functional materials, ion track and nanotechnology research. Metal thin film targets are usually prepared by physical vapor deposition methods, and the applicability of the deposition material and substrate is very wide. The physical vapor deposition process can be summarized into the following three stages:

(1) Emission of particles for coating from the source material;

(2) Transport of particles to the substrate;

(3) Condensation of particles on the substrate, nucleation, and growth into a film.

According to the way particles are emitted from the source material, there are several different types of physical vapor deposition, which are briefly introduced below.

Vacuum evaporation film deposition has the advantages of simple operation, high speed, and high efficiency (see Figure 7.5), but at the same time, the film formed has poor adhesion to the substrate, large grain size, low strength, and poor reproducibility of the process. According to the type of heat source used, the commonly used evaporation methods are shown in Table 7.1.

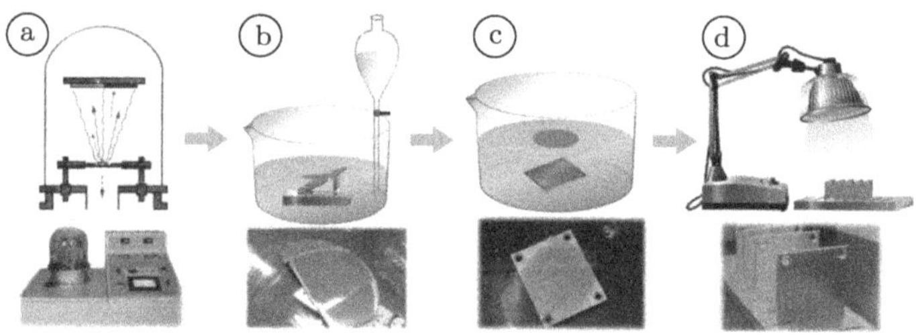

Fig. 7.5 Schematic diagram of vacuum evaporation coating and metal target preparation.

Table 7.1 Main methods and advantages and disadvantages of vacuum evaporation film.

Deposition Method	Advantages	Disadvantages
Resistance thermal evaporation: Heating resistor evaporates materials	Simple method, low cost, wide application	Can only deposit metals, unable to prepare high melting point metals, reacts with crucible
Electron beam evaporation: High energy electron beam bombardment	Can deposit active and high melting point materials; localized heating; can produce high purity films	Expensive and complex equipment
Laser evaporation: Laser bombardment of source materials	Localized surface heating, low thermal contamination; enables preparation of various materials	Material splashing, low efficiency, small film dimensions; high and complex laser requirements
Arc evaporation: Controlled plasma motion	High particle energy, high degree of film crystallization	Relatively low purity, requires special treatment

Sputter coating refers to the process of using particles with sufficient high energy to bombard the surface of a solid substrate, causing clusters of atoms to be emitted, which then deposit on the substrate to form a film. The angular distribution of the sputtered particles depends on the direction of the incident particles, and the sputtering rate depends on the mass and energy of the incident particles. Compared with vacuum evaporation, sputter coating has the following advantages:

(1) For the material to be coated, as long as it can be made into a disc-shaped substrate, sputtering can be achieved;

(2) The film obtained by sputtering has better adhesion to the substrate;

(3) The film obtained by sputtering has high purity and good density;

(4) The sputtering process has good reproducibility and controllability of film thickness, and can obtain a uniform film over a large area.

The disadvantages of sputter coating are that the deposition rate is relatively low, and the substrate may experience temperature rise due to irradiation by the plasma, which can cause significant loss of valuable materials such as gold and platinum. Since the sputtered particles have high energy, the resulting film generally has high internal stress, and when preparing thick self-supported films, it is prone to breakage.

The mainstream sputtering methods currently include direct current (DC) sputtering, radio frequency (RF) sputtering, and RF magnetron sputtering. The process of DC sputtering is as follows: the substrate to be coated is connected to the cathode of the power supply, and the substrate corresponding to the substrate is connected to the anode of the power supply. When a DC voltage of 1 to 5 kV (current density 1 to 10 mA/cm^3) is applied between the two electrodes, the inert gas such as argon (1.3 to 13 Pa) filled into the vacuum chamber will glow and discharge. The inert gas ions from the glow discharge plasma bombard the substrate, splashing out neutral atoms or clusters, which then deposit on the substrate to form a film. At the same time, the secondary electrons generated by the bombardment of the substrate participate in the glow discharge process, allowing the glow discharge to be self-sustained. A significant limitation of DC sputtering is that if the cathode substrate is non-conductive, positive charges will accumulate on the cathode and cannot be neutralized, eventually forming a reverse electric field that inhibits the normal sputtering process. Therefore, DC sputtering can only be used for sputtering conductive materials. To solve this problem, RF sputtering was developed. This method applies an RF alternating electric field to the cathode. The electrons oscillating in the alternating electric field have sufficient energy to produce ionizing collisions, thereby achieving self-sustained discharge. The RF sputtering method can greatly expand the variety of substrates.

It should be noted that since the RF field is applied between the two electrodes, electrons that escape from between the two electrodes after random collisions will not oscillate in the RF field. RF magnetron sputtering can effectively solve this problem. It increases the relaxation process of electrons in the plasma through an alternating electromagnetic field. At the same time, due to the different migration rates of electrons and ions in the RF field, the existence of cathode self-bias maintains the potential difference required for sputtering [3].

There are two methods to achieve self-supporting metal films. As shown in Figure 7.6, one is to evaporate the metal film on a substrate that can dissolve in a certain solution. By dissolving the substrate, the film can float on the liquid surface, thus achieving self-supporting. The other is to deposit a sacrificial layer between the substrate and the metal film. By dissolving the sacrificial layer, the film and the substrate can be separated, and then a self-supporting metal film can be obtained by retrieving it with a perforated plate.

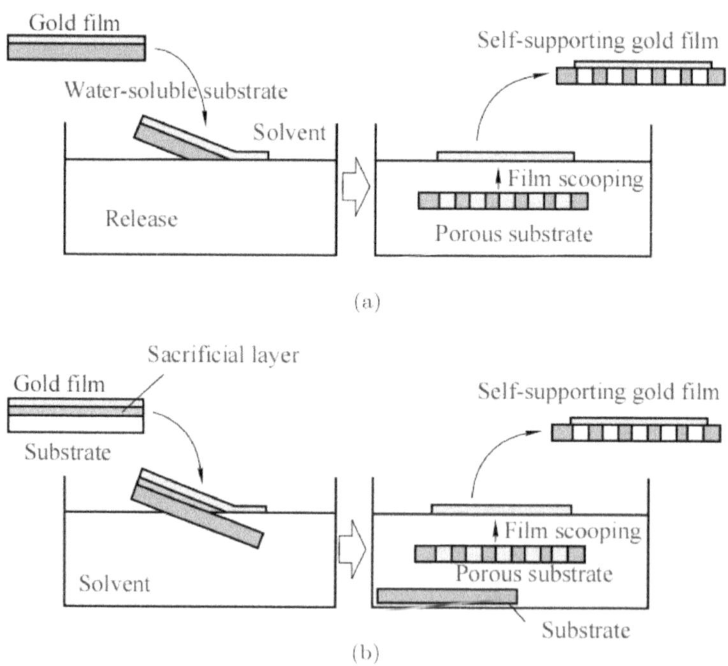

Fig. 7.6 Two self-supporting realization methods for self-supporting metal film targets [4].

Soluble substrates include Cu foil, Al foil, NaCl polished sheets, single-crystal silicon wafers, and some organic films (such as collodion), etc. For example, Cu foil can dissolve in ammonium trichloroacetate solution, Al foil can dissolve in NaOH solution, NaCl dissolves in water, glass and single-crystal silicon wafers dissolve in HF acid, and organic films can dissolve in organic solvents such as acetone. However, these soluble substrates all have certain defects to some extent. In practical applications, Cu foil and Al foil can cause metal diffusion, making the prepared film impure; NaCl polished sheets are prone to deliquescence, causing the film to blister; HF acid can dissolve the film at the same time when dissolving glass or single-crystal silicon wafers; and during the dissolution of organic substrates, the interface tension between liquid and gas can cause the film to rupture.

Coating or growing a soluble release agent on a solid polished surface (such as a polished silicon wafer or glass slide) and then depositing the film, followed by dissolving the release agent, is usually a more effective method. Common release agents and their corresponding solvents are shown in Table 7.2.

In addition to the methods mentioned above, there is also a traditional and efficient method for preparing metal sheets (films) — rolling. Rolling is a method in which a metal blank is passed multiple times through the gap between rotating rolls, and the thickness of the blank is reduced and the area is increased by the compression of the rolls. For metal targets used in laser acceleration, rolling can produce large-area metal films in large quantities and quickly, eliminating the de-molding step and directly obtaining self-supported targets. However, the thickness of the metal target prepared by rolling depends on the precise adjustment of the roll gap and the ductility of the material, so it is difficult to prepare metal films thinner than 5 μm. When the thickness is less than 10 μm, the metal film is very fragile and needs to be handled with special care. Materials suitable for rolling metal targets include Al, Cu, Ti, etc.

7.1.3 *Diamond-Like Carbon (DLC) Films*

Diamond-like carbon (DLC) films are amorphous pure carbon films composed of sp2 and sp3 hybridized bonds. DLC films have good hardness and mechanical strength, as well as excellent transparency and smoothness, and are widely used as ultra-thin protective coatings on magnetic hard disks and their read heads. As laser acceleration targets, DLC films, due to their good mechanical strength, can be made into ultra-thin self-supported films

Table 7.2 Defilm agents commonly used in the preparation of self-supported films and their corresponding solvents.

Name	Chemical Formula	Melting Point /°C	Boiling Point /°C	Cold Water Solubility /(g/100 ml)	Hot Water Solubility /(g/100 ml)	Other Solvents
Barium chloride	$BaCl_2$	925	1520	37.5	59	-
Potassium chloride	KCl	776	500	34.7	56.7	Glycerol, ether
Sodium chloride	NaCl	801	1413	35.7	39.12	Glycerol
Nickel chloride	$NiCl_2$	1001	973	64.2	87.6	Alcohol, ammonia
Zinc chloride	$ZnCl_2$	283	732	432	615	Alcohol; soluble in ether
Barium iodide	BaI_2	740	-	170	-	Alcohol
Cesium iodide	CsI_2	621	1280	44	160	Alcohol
Potassium iodide	KI	686	1330	127.5	208	Alcohol, ammonia, acetone
Sodium iodide	NaI	651	1304	184	302	Alcohol, acetone, glycerol
Alanine	$C_3H_7NO_2$	295	258	Very soluble	Very soluble	Alcohol
Betaine	$C_5H_{11}NO_2$	241	-	Very soluble	Very soluble	Alcohol
Formvar	$C_5H_8O_2$	110	-	Insoluble	Insoluble	Chloroform, Toluene
Dextrose	$CH_2OHCHCHOH_{40}$	150	-	Soluble	Very Soluble	Alcohol, acetone
Acetamide	$CH_3(CH_2)_{15}NH_2$	-	-	Insoluble	Insoluble	Hexane
Sucrose	$C_{12}H_{22}O_{11}$	185	-	Soluble	Very soluble	Alcohol
Nitro-cellulose	$C_{12}H_{16}O_6(NO_3)_4$	100	-	Insoluble	Insoluble	Acetone, Isoamyl acetate

(3 to 5 nm), which is crucial for the Radiation Pressure Acceleration (RPA) mechanism. Their low light absorption, good chemical stability, and impact resistance help maintain integrity under the action of laser prepulses. DLC films can be synthesized by various methods, such as sputter deposition, pulsed laser ablation, ion beam deposition, and plasma-enhanced chemical vapor deposition, among which sputter deposition has been introduced earlier, and the following two methods are briefly described here.

Ion beam deposition (IBD) is a thin film deposition process that uses an ion source to deposit or sputter the target material (metal or dielectric) onto a substrate to form a metal or dielectric film. Because the ion beam energy is determined and highly collimated, compared with other physical vapor deposition (PVD) techniques, ion beam deposition can more precisely control the thickness and deposit very dense, high-quality films.

Plasma-enhanced chemical vapor deposition (PECVD) technology is a film preparation technique that relies on glow discharge plasma to induce chemical reactions in gaseous substances containing film components, thereby realizing the growth of film materials. Compared with conventional chemical vapor deposition (CVD), PECVD effectively utilizes the reaction characteristics of non-equilibrium plasma, fundamentally changing the way energy is supplied to the reaction system. Generally speaking, the growth of film materials using PECVD technology mainly includes the following three basic processes:

(1) In the non-equilibrium plasma, electrons undergo primary reactions with the reaction gas, causing the reaction gas to decompose and form a mixture of ions and reactive species;

(2) Various reactive species diffuse and transport to the film growth surface, while secondary reactions occur between the reactants;

(3) The primary and secondary reaction products that reach the growth surface are adsorbed on the surface and react, accompanied by the release of gaseous molecular products.

In the glow discharge plasma, electrons accelerated by the external electric field typically have kinetic energies of around 10 eV or even higher. Therefore, inelastic collisions between high-energy electrons and reaction gas molecules can cause gas molecules to ionize or decompose, producing neutral atoms and molecular products. Positive ions, accelerated by the ion layer electric field, collide with the upper electrode; there is also a smaller ion layer electric field near the lower electrode where the substrate is placed, so the substrate also experiences a certain degree of ion bombardment. The chemically active species (mainly groups) that reach the substrate and are adsorbed there are highly reactive. They react with each other to form the film.

Ma Wenjun *et al.* [5] used filtered cathodic vacuum arc deposition (FCVA) to prepare DLC films with a thickness of less than 5 nm. Compared with other methods, the following advantages of FCVA are particularly suitable for preparing self-supported DLC target bodies:

(1) The proportion of sp3 bonds in the film is as high as 85%, possessing the best diamond-like properties, being hydrogen-free and highly transparent. Using low-deposition-rate pulsed arcs can produce continuous, robust, ultra-thin DLC films, as thin as a few atomic layers;

(2) By adjusting the substrate bias, the energy of the deposited carbon ions can be changed, allowing control of the properties of the deposited DLC film over a wide range of parameters;

(3) FCVA is essentially a low-temperature deposition technique, enabling the use of various release agents for the preparation of ultra-thin self-supported films.

As shown in Figure 7.7, the general process of FCVA is as follows: A highly ionized plasma is generated by a pulsed arc on the surface of a graphite cathode, with an average drift energy of 10 to 30 eV, and then accelerated towards the anode. A vertical magnetic duct is placed between the substrate and the anode to filter out large particles from the plasma. Finally, atomic clusters deposit on the substrate (such as a silicon wafer) to nucleate, grow, and extend into a film. To control the energy of carbon ions, a bias can be applied to the substrate. The deposition rate is controlled by the current and the arc pulse frequency. To minimize the internal stress of the DLC film, a low bias voltage (around 20 V) and a low deposition rate (around 0.3 nm/min) should be used. As previously introduced, to prepare the self-supported film required for laser acceleration experiments, a sacrificial layer is also needed on the substrate. The sacrificial layer currently generally uses a sodium chloride film. Studies have shown that the corrugated structure of the sodium chloride film makes the strength and film-forming properties of the DLC layer deposited on it better [5]. Overall, the DLC film has good flatness, with surface undulations not exceeding 30 nm.

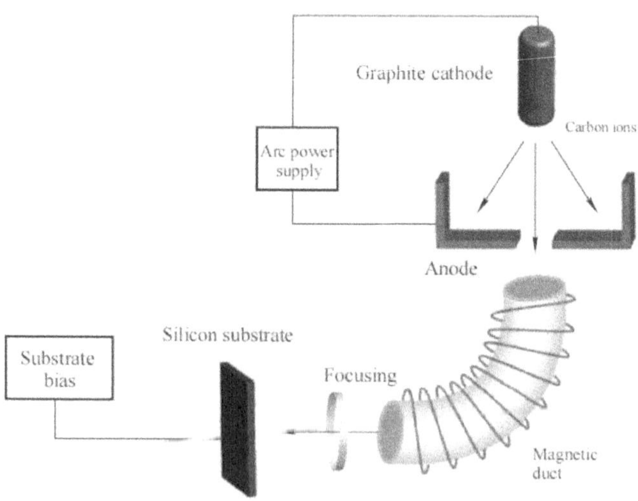

Fig. 7.7 Schematic diagram of filter cathode vacuum arc deposition (FCVA) [5].

7.2 Low-Density Porous Foam Targets

During the interaction between laser and target material, the maximum electron density of plasma that allows laser propagation is called the critical density, which is related to the laser wavelength. For a laser with a wavelength of 1 μm, the critical density is approximately $10^{21}/\text{cm}^3$. A target material with an electron density (n_e) much smaller than the critical density (n_c) is called a low-density target (such as a gas target), one much larger is called a high-density target (such as a solid target), and one close to the critical density is called a near-critical density target, as shown in Figure 7.8. The pre-ionization bulk density of near-critical density targets composed of light elements is in the range of 1 to 10 mg/cm^3. Foam materials can cover this density range, and they can be converted into near-critical density plasmas with steep boundaries after ionization in ultra-high light intensity, which is of great importance in laser acceleration.

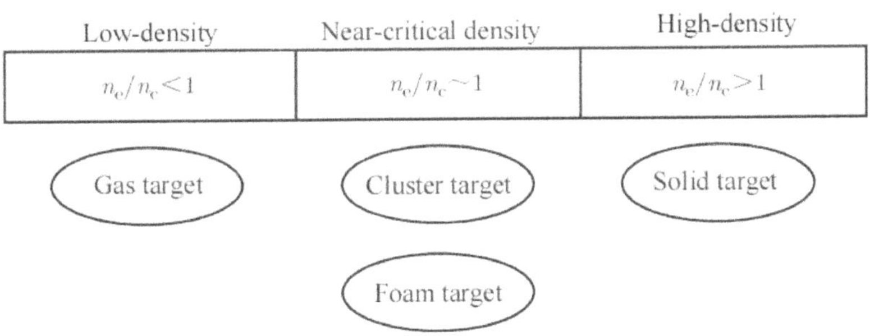

Fig. 7.8 Density distribution diagram of target material with different material states.

Porous foam is one of the most representative low-density foam materials. According to the pore size, they can be classified into microporous materials with pore sizes less than 2 nm, mesoporous materials with pore sizes between 2 and 50 nm, and macroporous materials with pore sizes greater than 50 nm. Currently, the pore sizes of prepared porous foam materials are mostly greater than 50 nm. The laser focal spot size in ion acceleration is approximately 3 μm $\times$ 3 μm. When low-density foam materials are used as targets, the smaller the pore size, the more uniform the plasma, which can suppress the development of plasma instabilities. It is very challenging to achieve a bulk density within the range of 1 to 10 mg/cm^3 while maintaining nanoscale uniformity in porous foam materials.

7.2.1 *Polymer Porous Foam Targets*

Due to their ease of molding and strong plasticity, polymer materials are one of the main raw materials for making foam materials. In laser plasma experiments, low-density foam targets are mainly prepared by chemical synthesis methods. The main preparation methods include sol-gel method, photochemical curing molding, freeze-drying, and electrospinning, etc. The commonly synthesized organic compounds include polyacrylates, polysaccharides, poly(3-methyl-1-pentene) (PMP), resorcinol formaldehyde (RF), etc., and Figure 7.9 summarizes the elemental composition, density, and pore diameter of commonly used polymer foam targets.

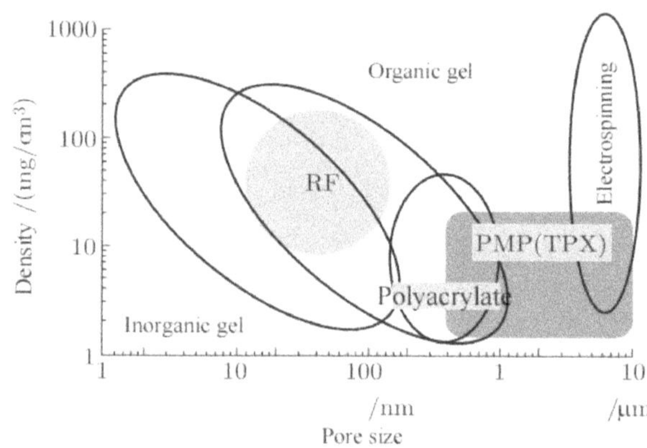

Fig. 7.9 Range of density and pore diameter of polymer foam materials [7].

The sol-gel method uses inorganic materials or metal alkoxides as precursors, which are uniformly mixed in the liquid phase and undergo hydrolysis and condensation chemical reactions to form a stable transparent sol gel system in the solution. The sol is aged, and the colloidal particles slowly polymerize to form a three-dimensional network structure of gel, with the space between the gel network filled with solvent that has lost its ???. The gel is then dried and sintered to prepare materials with molecular and even sub-nanometer structures, as shown in Figure 7.10.

Photochemical curing (photocuring) of foam materials involves irradiating a photosensitive solution of monomers and initiators with light of a certain wavelength (usually ultraviolet) to produce free radicals. Subsequently, the free radicals polymerize to form a uniform gel. Photocuring molding induces extensive rapid cross-linking with high cross-linking

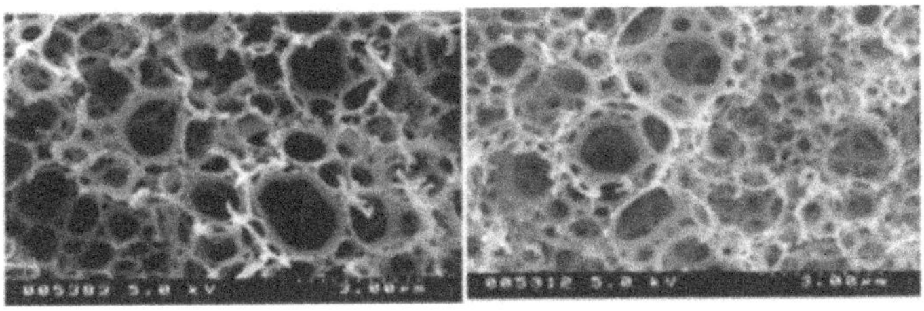

Fig. 7.10 Microstructure of porous foam prepared by sol-gel method [7].

density. Compared with the sol-gel method, the resulting foam materials have smaller pore sizes and shrinkage rates. Both photocuring molding and sol-gel method form porous structures through the contraction and drying of polymer emulsions, with relatively uniform pore distribution and pore size. Their liquid precursors can be used to cast complex-shaped components. These polymer foams themselves can also serve as molds, allowing the casting of non-metallic solutions to form shapes, followed by high-temperature pyrolysis to remove the polymer layer, yielding non-metallic foam materials.

Electrospinning, with its simple manufacturing apparatus, low cost, wide range of spinnable materials, and controllable process, has become an important route for preparing nanofiber materials. Specifically, a polymer solution or melt is first charged with several thousand to tens of thousands of volts of high-voltage static electricity. The charged polymer droplet is accelerated at the Taylor cone vertex of the capillary under the action of the electric field force. When the electric field force is strong enough, the polymer droplet overcomes the surface tension to form a jet. During the jetting process, the solvent evaporates or solidifies, and finally, it lands on the receiving device, forming a fibrous mat similar to nonwoven fabric, as shown in Figure 7.11. Using electrospinning, low-density materials with densities in the range of tens to hundreds of mg/cm^3 can be prepared.

7.2.2 *Carbon Foam Low-Density Targets*

Carbon atoms can bond to form a variety of materials with different morphologies, such as graphene, carbon nanotubes, porous carbon, and diamond, through sp2 and sp3 bonds. Low-density pure carbon foam materials can serve as excellent near-critical density targets. Ma Wenjun *et al.* were the first in the world to use carbon nanotubes to prepare near-critical density targets for laser acceleration experiments, significantly enhancing

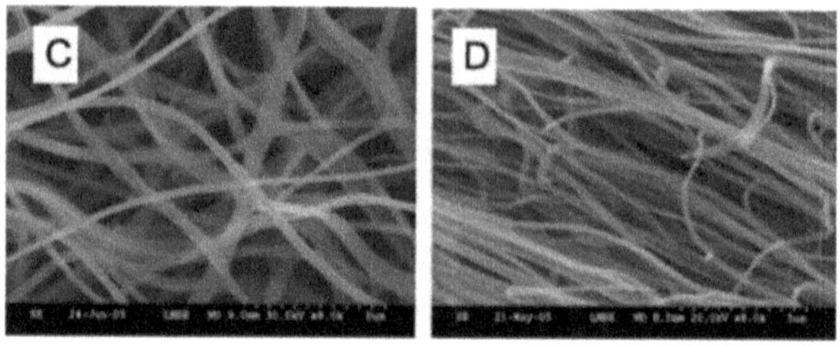

Fig. 7.11 Multi-air gap low-density film prepared by electrospinning method (SEM image) [7].

the acceleration energies of carbon and gold ions [8]. The main methods for preparing carbon nanotubes currently include arc discharge, laser ablation, and chemical vapor deposition. Arc discharge uses high-purity graphite as electrodes, and a high voltage is applied to ignite an arc in a vacuum. Under high-temperature conditions, carbon atoms restructure and grow into carbon nanotubes. This method produces carbon nanotubes with high reaction temperatures, the highest degree of carbon atom crystallization, high purity, and perfect structure. However, the resulting carbon nanotubes are in the form of particles and powders, making it difficult to form films and thus not suitable as targets for laser acceleration experiments. Laser ablation has low yield, limited area, short carbon nanotube growth time, and is difficult to control the fine structure. Chemical vapor deposition is the main method for large-scale preparation of carbon nanotube films. The following introduces the method of preparing carbon nanotube films using floating catalyst chemical vapor deposition.

The basic process is shown in Figure 7.12. A mixture of carbon sources (e.g., methane) is uniformly mixed and introduced into a quartz tube that has been evacuated of air, while carrying catalysts that have been sublimated by heating. In the central high-temperature reaction zone of the tube, the carbon atoms in the reaction gas are ?? by the catalyst, and carbon nanotubes grow on the surface of the nanocatalyst, floating in the carrier gas and moving towards the tail of the furnace with the gas flow. As the temperature at the tail of the furnace decreases, the carbon nan otubes interconnect and form a carbon nanotube film on the substrate. In the resulting carbon nanotube film, the carbon nanotubes are uniformly and randomly arranged on a microscopic scale (see Figure 7.13), forming a three-dimensional network with a bulk density as low as 1 mg/cm^3.

In addition, the Laboratory of Micro and Nano Structures at the

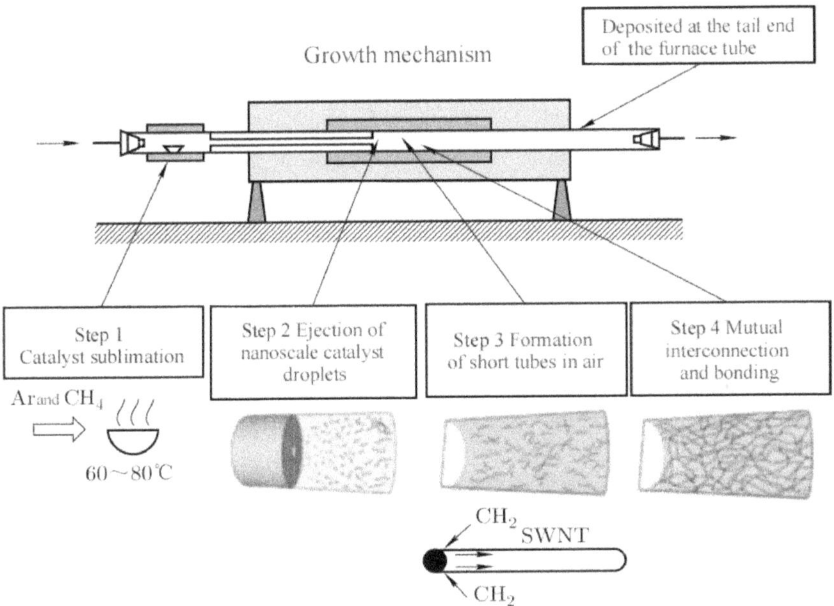

Fig. 7.12 Schematic diagram of carbon nanotube thin films prepared by floating catalytic chemical vapor deposition [8].

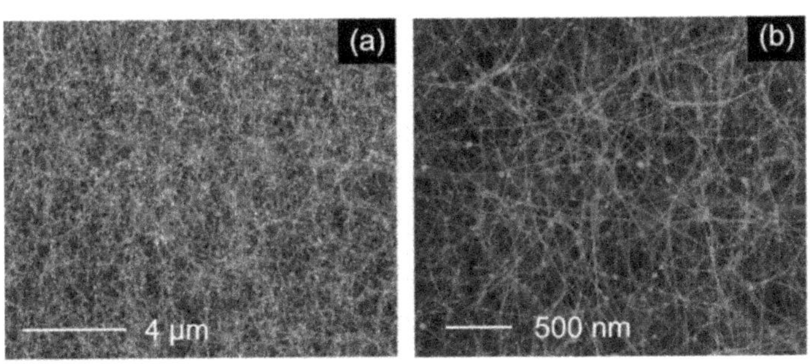

Fig. 7.13 Microstructure of carbon nanotubes [31] (SEM).

Polytechnic University of Milan used pulsed laser deposition (PLD) to prepare porous carbon foam materials [9]. The basic process is to use the second harmonic (532 nm) of an Nd: YAG laser (pulse duration 5 to 7 ns, repetition frequency 10 Hz, transmitted energy 0.8 J/cm^2) to ablate a pyrolytic graphite target in a chamber filled with argon gas. The surface layer of the graphite target expands rapidly due to heat and detaches from the base layer. The cluster particles collide with each other and deposit on the substrate to form porous carbon foam material (see Figure 7.14). The argon gas in the chamber increases the collision frequency between the ablated particles, causing carbon atoms to form clusters and nanoparticles in the plasma plume, resulting in a porous morphology of the assembled film.

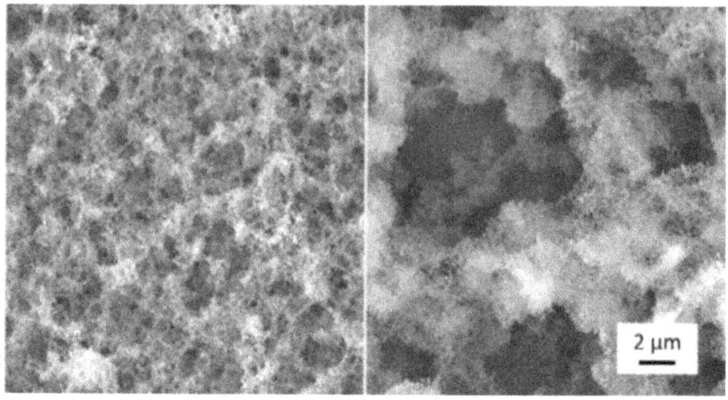

Fig. 7.14 SEM image of porous carbon foam target [10].

The pressure of argon gas in the chamber and the distance between the graphite substrate and the deposition substrate affect the parameters of the porous carbon foam target. By adjusting the parameters, the density and morphology of the target material can be changed [11]. When the gas pressure increases and the deposition distance is extended, the collision frequency of the expanding carbon nanoparticles with argon gas is higher, and the relaxation time and path are longer, resulting in a more uniform porous foam target with a lower average density.

The thickness gradient of the deposited material by pulsed laser deposition is relatively large, and static deposition can only produce a uniform area of about 1 cm^2. The uniformity can be improved by translating and rotating the substrate.

7.2.3 *Metal Porous Foam Targets*

The preparation method of metal foam targets is similar to that of porous carbon foam targets. First, metal cluster particles are detached from a high-purity metal substrate using a physical method. After colliding and relaxing with the gas in the chamber, they deposit on the substrate to form a high-density target material (see Figure 6 in Ref. [12]). The main preparation methods currently include vacuum resistance thermal evaporation, electron beam evaporation, and ion beam evaporation. Similarly, the density and uniformity of the target material can be controlled by adjusting the pressure of argon gas in the chamber and the deposition distance. The thickness of the target material can be controlled by adjusting the deposition time.

7.3 Microstructured Targets

Studies have shown that using targets with microstructures on the surface to interact with laser can increase the laser penetration depth, enhance the coupling degree with the plasma, leading to higher energy absorption rates and better laser acceleration effects. Currently, common microstructured targets include polymer microsphere targets, groove array targets, nanowire array targets, etc. [13, 14], which are briefly introduced below.

The preparation method of a microsphere target covered with a single layer of polymer microspheres is as follows. First, polystyrene microspheres with diameters in the range of 100 to 1000 nm, purchased from vendors, are dispersed in a mixture of water and ethanol (usually in a 1:1 ratio). Then, using a glass pipette with a curved tip, the microspheres are carefully coated on the water surface in a petri dish. The self-assembly of polystyrene microspheres at the water/air interface forms a closely packed single layer of microspheres. At this point, the solid film target material is immersed under the monolayer, and then slowly lifted, allowing the single microsphere layer to remain on the target material surface, forming a microstructured target with a periodic structure. Its surface morphology is shown in Figure 7.15.

Fig. 7.15 SEM image of polystyrene microsphere structure target [13].

The method for preparing groove arrays (see Figure 7.16 for morphology) on a planar target surface includes photolithography and nanoimprint lithography.

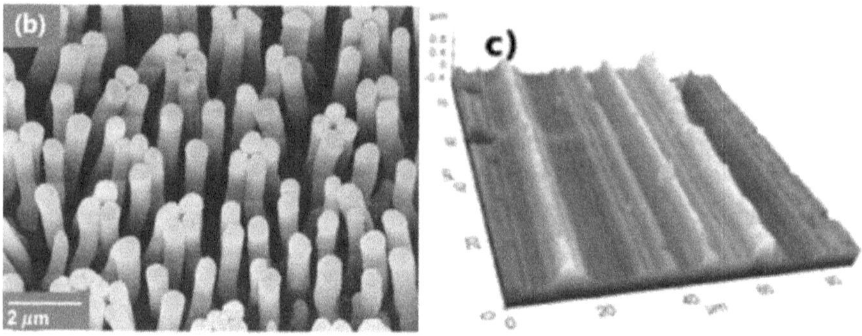

Fig. 7.16 SEM images of nanowire array target (left) and groove array microstructure target (right) [15].

The general process of photolithography is to spin-coat a layer of photoresist on the surface of the target material to be etched, followed by exposure using a mask. After exposure, the sample is placed in a developer to develop the pattern. Depending on whether positive or negative photoresist is used, the corresponding photoresist is removed to form the groove array pattern. The target material is then etched. There are two main etching methods: chemical etching and physical etching. Chemical etching involves using a chemical solution to react with and corrode the material outside the photoresist film, removing the part not covered by the film to form groove arrays. Physical etching mainly involves ion beam or electron beam bombardment of the target material to etch away the material in the area not covered by the photoresist film, forming groove arrays. After the groove arrays are formed on the target material surface, the photoresist film is removed to obtain the final target.

Another method for preparing groove array microstructures is nanoimprint lithography, which is generally used for polymer materials. Nanoimprint lithography is a low-cost, high-resolution, and high-yield fabrication method. The general process involves first coating a polymer layer (such as PMMA), and then pressing a hard mold with the desired pattern into the polymer layer under certain temperature and pressure conditions. The mold is removed after the polymer layer is softened. Any residual adhesive can be removed by reactive ion etching.

The temperature for high-quality nanoimprint lithography is generally above the glass transition temperature (T_g) of the polymer, with a pressure between 5 and 10 MPa, depending on the softening degree of the polymer layer. The pressure applied to the mold is usually removed when the temperature is lowered to about 10°C above the glass transition temperature. The mold is pulled out when the polymer layer is cooled to below 50°C below the glass transition temperature. The realization of high-quality nanoimprint lithography depends on the mold, the imprint layer material, the demolding technique, and the alignment accuracy.

Polymer nanowire array targets can be prepared using anodic aluminum oxide (AAO) templates. AAO templates have numerous porous channels, and the pore size and depth can be selected according to needs. The preparation process involves first baking the polymer to soften it, then closely attaching it to the surface of the AAO template and applying pressure. The polymer is then heated to above its glass transition temperature. The softened polymer enters the AAO template under pressure, forming nanowires in the porous channels. Finally, the AAO template is chemically etched away to obtain the nanowire array target [16–18] (see Figure 7.16).

7.4 Liquid Thin Film Targets

Laser ion acceleration generally produces an exponential distribution of ion energy spectra. The number of high-energy ions in a single shot is relatively small. To achieve the average beam current required for certain applications, a higher repetition rate of target shooting is needed. The aforementioned target materials all have a common problem when facing applications: the target material is damaged after being shot, making it difficult to achieve high-repetition-rate target shooting.

A liquid in a flowing state can form a film with a thickness in the μm to nm range, which can exist independently without a substrate or container. The liquid molecules in the film are constantly renewed, and the film can quickly return to its initial state after being disturbed or damaged. Using such a film as a target material for laser acceleration has the significant advantage of being able to quickly recover after being shot, allowing for the next shot in high-repetition-rate target shooting. The preparation method of liquid thin film targets usually involves applying a driving force to the liquid in a certain direction to spread it into a film, which then closes under the action of surface tension. The following briefly introduces the process of preparing liquid thin film targets using the dual-column collision method.

The basic process of the dual-column collision method is as follows: Two high-speed jets are sprayed from two cylindrical tubes with diameters of about tens of μm, and they collide at a certain angle. The momentum in the opposite direction cancels out, forming a radial momentum in the perpendicular direction, and a closed liquid film is formed under the action of surface tension, as shown in Figure 7.17 [19]. Liquid thin film targets are prepared and used in real-time, so it is necessary to know the properties of the target material prepared under certain conditions in advance. Researchers have found that for the dual-column collision method: the smaller the jet diameter, the thinner the liquid film, and the larger the collision angle, the thinner the liquid film and the smaller the length-to-width ratio [20]; high-viscosity fluids make the film thicker and more uniform [21]; when the two liquid columns are not colliding centrally, the liquid film is thinner and has a certain deflection angle [22].

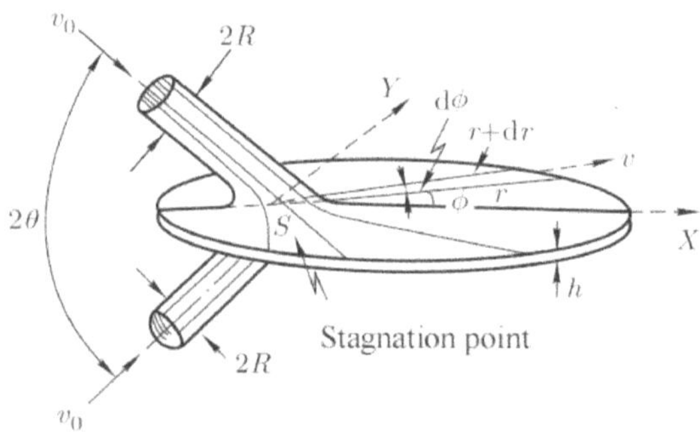

Fig. 7.17 Physical model of liquid column collision [19].

In addition to the dual-column collision method, another method for preparing liquid thin film targets is to use a converging nozzle, which is similar in principle to the dual-column collision method. The difference is that the "collision" process takes place inside the nozzle. In 2017, Gallinis *et al.* used a precision square converging nozzle made by two-photon 3D printing to obtain a liquid film with a thickness of about 1 μm [23]. In 2018, Ha *et al.* quantitatively studied the dependence of liquid film size and thickness on length-to-width ratio, convergence angle, and fluid properties for square converging nozzles [24]. Compared with the dual-column collision method, the converging nozzle method has a simpler device, no need for alignment, stable state, and higher portability. However, it has lower degrees of freedom for adjustment and can only control the liquid film by changing the flow rate.

Currently, research on liquid thin film targets is still in its infancy, with only a few works conducting laser ion acceleration experiments based on liquid thin film targets [25, 26]. The preparation of ultra-thin, large uniform liquid films is a challenge that needs to be overcome in the future development of liquid thin film targets. Improving the stability and vacuum compatibility of liquid films is also an important technical issue. Establishing an online film thickness and stability diagnostic system will facilitate real-time feedback control of the liquid film. With the progress and improvement of these technologies, liquid films, due to their rapid renewal capability in high-repetition-rate applications, material recyclability in terms of economy, and flexibility in adjustment, are expected to replace solid targets and become the mainstream target material in laser-driven ion sources.

References

[1] Zhao Junli. Study on the preparation and film formation mechanism of polymer thin films by dip coating [D]. Shanghai: Donghua University, 2014.

[2] Seuferling S., Haug M. A. O., Hilz P., *et al.* Efficient offline production of freestanding thin plastic foils for laser-driven ion sources [J]. *High Power Laser Science and Engineering*, 2017, 5: e8.

[3] Zheng Weitao. Thin Film Materials and Thin Film Technology [M]. Beijing: Chemical Industry Press, 2004.

[4] Miyamoto Y., Fujii Y., Yamano M., *et al.* Preparation of self-supporting Au thin films on perforated substrate by releasing from water-soluble sacrificial layer [J]. *Japanese Journal of Applied Physics*, 2016, 55: 07LE05.

[5] Ma W., Liechtenstein V. K., Szerypo J., *et al.* Preparation of self-supporting diamond like carbon nanofoils with thickness less than 5 nm for laser-driven ion acceleration [J]. *Nuclear Instruments & Methods in Physics Research*, 2011, 655(1): 53–56.

[6] Harigai T., Miyamoto Y., Yamano M., *et al.* Self-supporting tetrahedral amorphous carbon films consisting of multilayered structure prepared using filtered arc deposition [J]. *Thin Solid Films*, 2019, 675(APR.1): 123–127.

[7] Nagai K., Musgrave C. S. A., Nazarov W. A review of low density porous materials used in laser plasma experiments [J]. *Physics of Plasmas*, 2018, 25(3): 030501.

[8] Szerypo J., Ma W., Bothmann G., *et al.* Target fabrication for laser-ion acceleration research at the Technological Laboratory of the LMU Munich [J]. *Matter and Radiation at Extremes*, 2019, 4(3): 035201.

[9] Zani A., Dellasega D., Russo V., *et al.* Ultra-low density carbon foams produced by pulsed laser deposition [J]. *Carbon*, 2013, 56: 358–365.

[10] Prencipe J., Spattoni A., Dellasega D., *et al.* Development of foam-based layered targets for laser-driven ion beam production [J]. *Plasma Physics and Controlled Fusion*, 2016, 58(3): 034019.

[11] Prencipe I., Dellasega D., Zani A., *et al.* Energy dispersive X-ray spectroscopy for nanostructured thin film density evaluation [J]. *Science and Technology of Advanced Materials*, 2015, 16(2): 025007.

[12] Akimova I. V., Akunets A. A., Borisenko L. A., *et al.* Metals produced as nano-snow layers for converters of laser light into X-ray for indirect targets and as intensive EUV sources [J]. *Journal of Radioanalytical & Nuclear Chemistry*, 2014, 299(2): 955–960.

[13] Klimo O., Psikal J., Limpouch J., *et al.* Short pulse laser interaction with microstructured targets: simulations of laser absorption and ion acceleration [J]. *New Journal of Physics*, 2011, 13(5): 407–470.

[14] Klimo O., Psikal J., Limpouch J., *et al.* Simulations of short pulses laser interaction with targets having a submicron surface structure: energy absorption and ion acceleration [C]. Laser Acceleration of Electrons, Protons, and Ions and Medical Applications of Laser-Generated Secondary Sources of Radiation and Particles. SPIE, 2011, 8079: 103–110.

[15] Ceccotti T., Floquet V., Sgattoni A., *et al.* Evidence of resonant surface-wave excitation in the relativistic regime through measurements of proton acceleration from grating targets [J]. *Physical Review Letters*, 2013, 111(18): 185001.

[16] Passoni M., Sgattoni A., Prencipe I., *et al.* Toward high-energy laser-driven ion beams: nanostructured double-layer targets [J]. *Physical Review Accelerators and Beams*, 2016, 19(6): 061301.

[17] Jiang S., Ji L. L., Audesirk H., *et al.* Microengineering laser plasma interactions at relativistic intensities [J]. *Physical Review Letters*, 2016, 116(8): 085002.

[18] Dimitri, Khaghan, Mathieu, *et al.* Enhancing laser-driven proton acceleration by using micro-pillar arrays at high drive energy [J]. *Scientific Reports*, 2017, 7(1): 11366.

[19] Taylor G. The dynamics of thin sheets of fluid [J]. *Proceedings of the Royal Society of London*, 1959: 289.

[20] Ibrahim E. A., Przekwas A. J. Impinging jets atomization [J]. *Physics of Fluids A: Fluid Dynamics*, 1991, 3(12): 2981–2987.

[21] Lu J., Corvalan C. M. Influence of viscosity on the impingement of laminar liquid jets [J]. *Chemical Engineering Science*, 2014, 119: 182–186.

[22] Panao M. R. O., Delgado J. M. D. Effect of pre-impingement length and misalignment in the hydrodynamics of multijet impingement atomization [J]. *Physics of Fluids*, 2013, 25(1): 012105.

[23] Galinis G., Strucka J., Barnard J. C. T., *et al.* Micrometer-thickness liquid sheet jets flowing in vacuum [J]. *Review of Scientific Instruments*, 2017, 88(8): 083117.

[24] Ha B., DePonte D. P., Santiago J. G. Device design and flow scaling for liquid sheet jets [J]. *Physical Review Fluids*, 2018, 3(11): 114202.

[25] Morrison J. T., Feister S., Frische K. D., *et al.* MeV proton acceleration at kHz repetition rate from ultra-intense laser liquid interaction [J]. *New Journal of Physics*, 2018, 20(2): 022001.

[26] Puyuelo-Valdes P., de Luis D., Hernandez J., *et al.* Implementation of a thin, flat water target capable of high-repetition-rate MeV-range proton acceleration in a high-power laser at the CLPU [J]. *Plasma Physics and Controlled Fusion*, 2022, 64(5): 054003.

[27] `https://www.ossila.com/pages/spin-coating#introduction-to-spin-coating` [Z].

[28] Dalí G. C., Lagares D. T. Nanobiomaterials in hard tissue engineering: applications of nanobiomaterials [M]. Oxford: William Andrew Publishing, 2016: 1–31.

[29] Zhao Junli. Study on the preparation and film formation mechanism of polymer thin films by dip coating [D]. Shanghai: Donghua University, 2014.

[30] Wang Pengjie. Research on heavy ion acceleration in ultra-short and ultra-strong laser interaction with nano targets [D]. Beijing: Peking University, 2022.

[31] Wang P., Qi G., Pan Z., *et al.* Fabrication of large-area uniform carbon nanotube foams as near-critical-density targets for laser-plasma experiments [J]. *High Power Laser Science and Engineering*, 2021, 9: e29.

Chapter 8

Laser Ion Accelerators and Diagnostics

A typical laser ion accelerator usually includes: a high-power laser system, a laser quality enhancement system, an ion acceleration system, a diagnostic system, and a control system. Among these, the high-power laser system is used to provide high-power femtosecond laser pulses; the laser quality enhancement system is used to improve the temporal contrast and laser wavefront quality of the laser beam; the ion acceleration system is used for ion acceleration driven by ultra-intense lasers; the diagnostic system is used to obtain information on the laser and beam during and after acceleration; and the control system is used to maintain the efficient and orderly operation of the accelerator. Figure 8.1 shows a typical laser ion accelerator device [1]. This chapter will mainly introduce the laser quality enhancement system, the ion acceleration system, and the diagnostic system.

8.1 Laser Quality Enhancement System

The primary function of the laser quality enhancement system is to improve the stability, focusability, and temporal contrast of laser pulses. It consists of an optical path self-alignment and monitoring system, an adaptive optics (AO) system, and a plasma mirror system. The main function of the optical path self-alignment and laser monitoring is to ensure the stable transmission of laser pulses from the high-power laser system to the ion acceleration system. The optical path self-alignment system, in conjunction with the laser monitoring system, can achieve real-time alignment of the optical path. The following sections focus on introducing the adaptive optics system and the plasma mirror system.

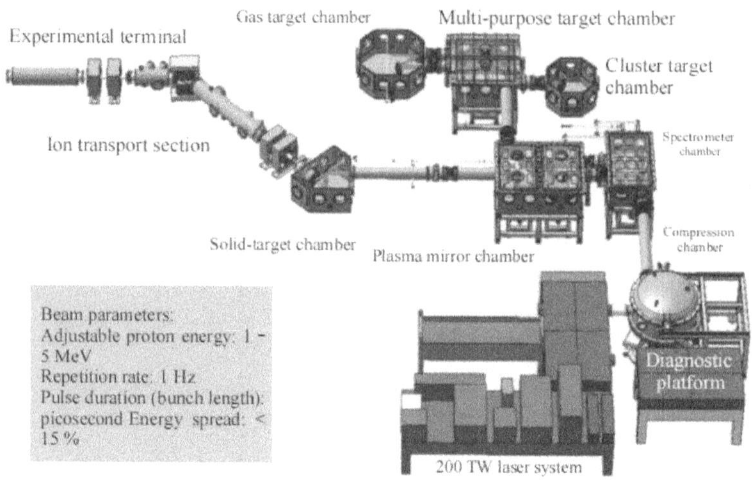

Fig. 8.1 Laser ion accelerator at Peking University.

8.1.1 *Adaptive Optics (AO) System*

Given a constant laser energy and pulse width, the laser focal spot determines the laser power density, while the laser wavefront phase determines the quality of the focal spot. Both factors will affect the final results of ion acceleration experiments. During the transmission of the optical path, optical elements such as mirrors, lenses, waveplates, and polarizers can cause wavefront phase distortions. To enhance the power density of the output pulse, it is necessary to use an adaptive optics (Adaptive Optics, AO) wavefront correction system to correct these distortions. The AO wavefront correction system is a feedback loop system that integrates measurement, control, and correction. It mainly includes: a wavefront sensor, a wavefront processor, and a wavefront corrector.

The wavefront sensor is used to measure the phase of the laser wavefront. There are currently two main methods for measuring the phase of a laser wavefront: one is the Shack–Hartmann method, which analyzes the shape of the wavefront and the distribution of intensity by measuring the position and intensity of the spots after the micro-lens array, and then reconstructs the phase of the laser beam using Zernike polynomials [18] or a zonal method; the other is the shearing interferometry method [19], which measures the phase of the laser beam using the principle of interference with a shearing interferometer. The principle of shearing interferometry is as follows: when the wavefront to be measured passes through the wavefront analyzer, a replicated wave with a certain lateral displacement from the measured wave is obtained through a specially designed grating. This replicated wave interferes with the measured wave, forming a lateral shearing interference, and interference fringes appear at the overlapping parts of the two waves. The measured wavefront may be a plane wave or a converging wave: for plane lateral shearing interference, a replicated wave is generated by a small lateral displacement of the measured wavefront within its own plane; while for converging lateral shearing interference, the replicated wave is generated by rotating the converging wave around its curvature center. The interference fringes contain differential information of the original wavefront, and the original wavefront can be reconstructed by specific analysis and quantitative calculation (inverse Fourier transform).

The wavefront corrector is used to compensate for wavefront phase distortions, a process carried out by a wavefront shifting device, primarily achieved by altering the refractive index or the optical path length. In laser ion accelerators, a deformable mirror is generally used as the wavefront corrector. The feedback loop of the wavefront correction system is shown in Figure 8.2. The wavefront sensor measures the current wavefront phase, which is then analyzed by the wavefront processor in comparison with the ideal phase. The correction amount is subsequently applied to the wavefront corrector, forming a cycle.

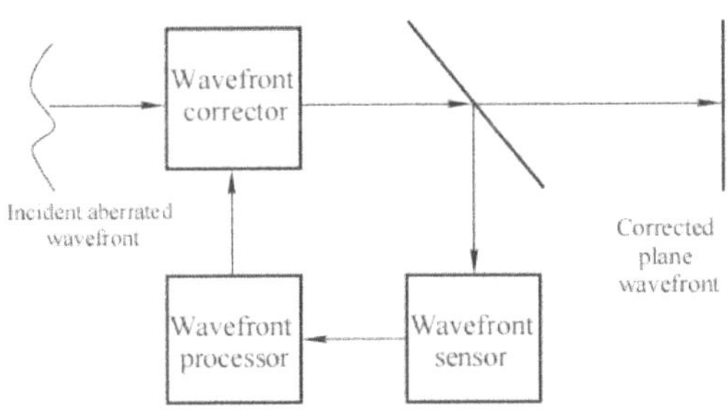

Fig. 8.2 Schematic diagram of the feedback loop of the wavefront correction system.

Under normal circumstances, high-power laser systems perform wavefront phase correction at the output stage. However, during laser transmission, optical elements introduce new wavefront distortions, reducing the focusing quality of the laser. Therefore, wavefront phase correction

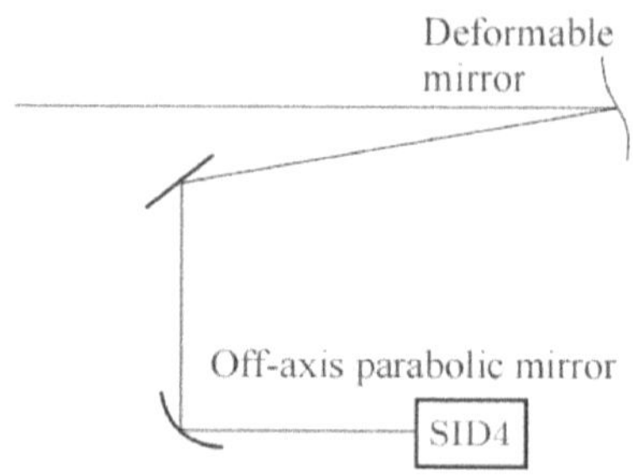

Fig. 8.3 Schematic diagram of wavefront correction (SID4 is a wavefront sensor based on shearing interferometry).

is still necessary within the laser quality enhancement system. A common wavefront phase correction setup is shown in Figure 8.3. The wavefront phase of the laser is measured after focusing, and then the phase is compensated by optimizing the deformable mirror to achieve the best laser focus.

8.1.2 *Plasma Mirror System*

With the advent of the PW (petawatt) laser era, laser facilities around the world are expected to achieve peak intensities greater than 10^{22} W/cm^2 in focal spots with diameters of a few μm. However, the ionization damage threshold of the target is typically less than 10^{13} W/cm^2. This means that under such extreme conditions, the laser temporal contrast must reach at least 10^9, otherwise the target material will be ionized and pre-expanded before the arrival of the main pulse. Information on laser prepulses was introduced in Section 5.1. To meet the ultra-high contrast requirement, the plasma mirror system has been introduced to further enhance laser contrast. Nearly 20 years of research has shown that plasma mirror technology is an effective method to improve laser contrast, and it has achieved good results in many experiments. Figure 8.4 shows the dual plasma mirror system of the PW laser at CoReLS in South Korea (upper figure) and its experimental effect on optimizing laser contrast (lower figure) [2].

The principle of plasma mirror technology is shown in Figure 8.5. The plasma mirror is a reflective mirror made of quartz or glass coated with an anti-reflection coating, with high requirements for the flatness of the mirror surface. In experiments, by switching between the low reflectivity state of the anti-reflection mirror and the high reflectivity state of the plasma, a

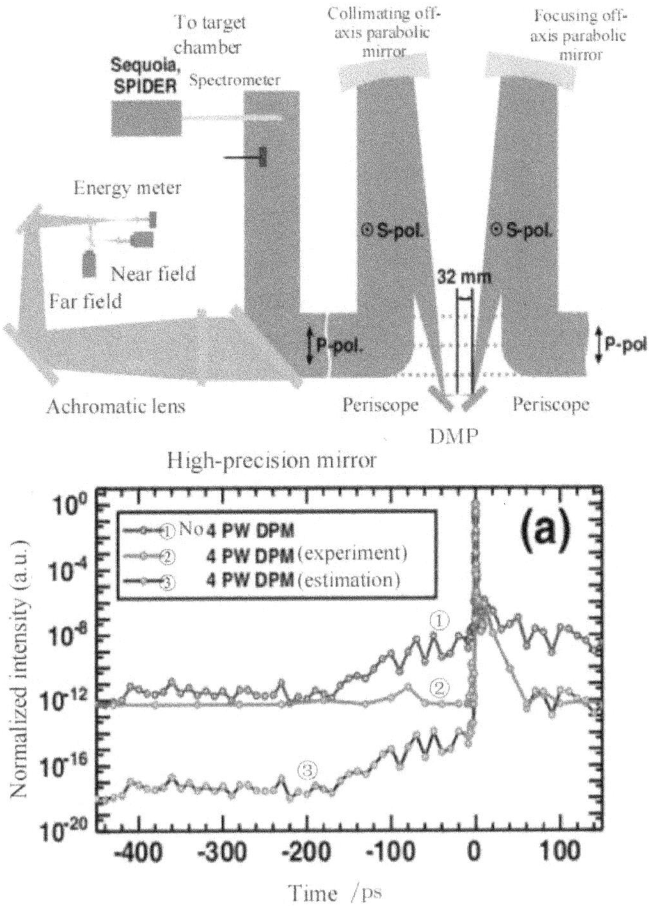

Fig. 8.4 Dual plasma mirror system for PW Laser at CoReLS, South Korea [2]. Sequola, SPIDER: A pulse width measurement device; DMP: Dual plasma mirror.

temporal switch for the laser pulse is formed, thereby enhancing the pulse contrast. The laser is focused by a long focal length off-axis parabolic mirror, causing the focal point to fall on the surface of the anti-reflection mirror either before or after the focus. The spontaneous amplified emission (ASE) and prepulse of the laser pulse front have low intensity and can be transmitted through the surface of the plasma mirror. When the main laser pulse arrives, the laser intensity rapidly increases to the ionization threshold of the mirror material, thus ionizing and producing plasma. As the plasma density rapidly rises to the critical density, the laser is completely reflected

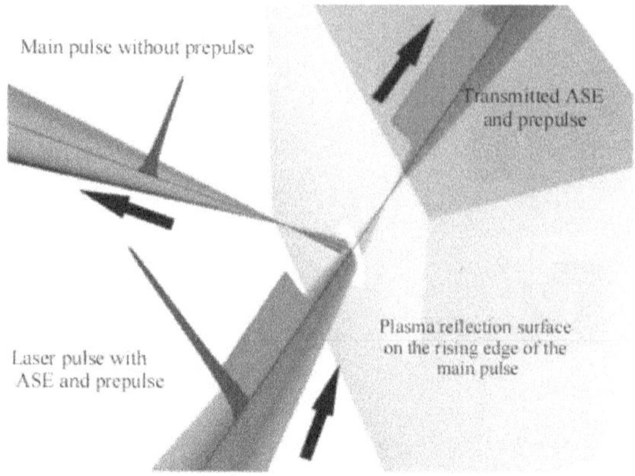

Fig. 8.5 Schematic diagram of the principle of plasma mirror technology [3].

at the critical density layer. In this way, the low-intensity laser pulse front is mainly transmitted through the mirror surface, while the high-intensity main laser pulse is reflected by the mirror surface. The contrast of the laser pulse reflected by the plasma mirror is significantly improved. Using this plasma technology optical switch, the laser contrast can typically be increased by two orders of magnitude [3]. If a dual plasma mirror system is used, the laser contrast can be further enhanced.

Further research has shown that plasma mirror technology remains effective on the picosecond scale [4]. When interacting with an ultra-intense laser, the triggering of the mirror surface plasma is completed within a few ps. Thus, the switching of the plasma mirror's reflective state also occurs on the picosecond scale, enabling the technology to optimize picosecond-scale pulse contrast (Figure 8.6). This is difficult to achieve with other techniques [5]. In practical applications, when the laser itself has good contrast and beam uniformity, the plasma mirror can achieve excellent results. The quality of the beam after reflection remains essentially unchanged, although there may be a slight blue shift in the central wavelength, but the pulse width change is minimal. In practical applications, it is necessary to verify the beam quality after plasma mirror reflection to ensure that the reflected laser still has sufficient beam quality and far-field focusing performance.

In summary, the plasma mirror system can enhance the contrast of ultrashort and ultra-intense laser pulses by 2 to 5 orders of magnitude in practical applications, while maintaining a high reflectivity that ensures most of the laser energy is transmitted to the target. Therefore, with the aid of the plasma mirror system, laser contrast can be significantly improved,

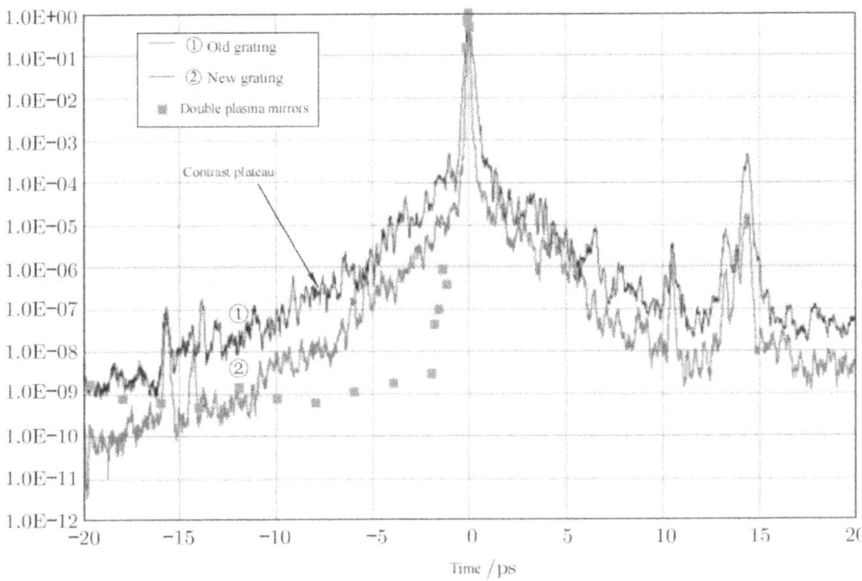

Fig. 8.6 Significant enhancement of pulse contrast on the picosecond scale through a plasma mirror. Although optimized gratings can improve pulse contrast to some extent, the slow rising edge on the picosecond scale is still difficult to eliminate [5].

which helps to achieve more efficient radiation pressure acceleration, resulting in proton beams with higher cut-off energies and smaller divergence angles. The use of plasma mirror technology in the future holds significant importance in the fields of laser ion acceleration and high-brightness radiation.

8.2 Ion Acceleration System

The ion acceleration system mainly consists of six parts: focusing system, focal spot monitoring system, target system, vacuum system, plasma parameter diagnostic system, and ion diagnostic system. Figure 8.7 shows a common layout of the ion acceleration system. High-power laser pulses are focused and incident on the surface of a solid target at a certain angle to interact with the target, thereby achieving ion acceleration.

(1) Focusing System

The six-dimensional focusing system is composed of an off-axis parabolic reflector (Off Axis Parabola, OAP) and a six-dimensional adjustment system, as shown in Figure 8.8(a). The six-dimensional adjustment system

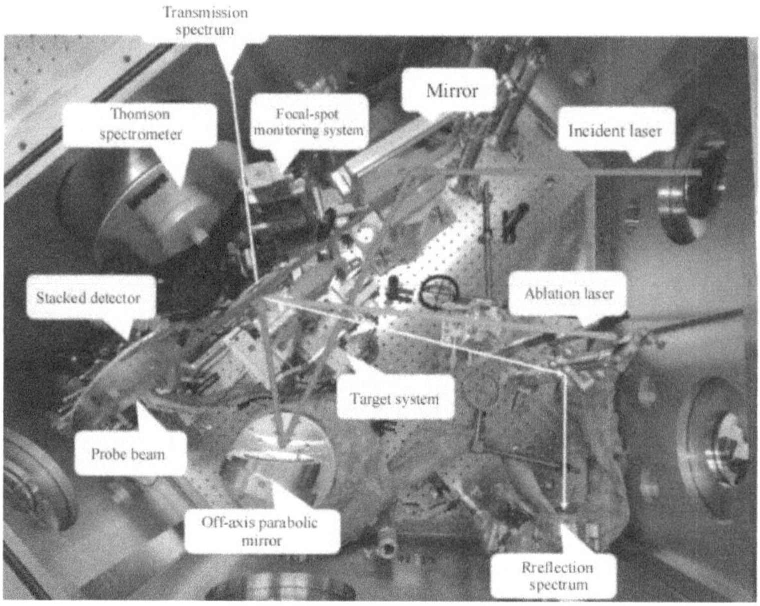

Fig. 8.7 Schematic diagram of the ion acceleration system.

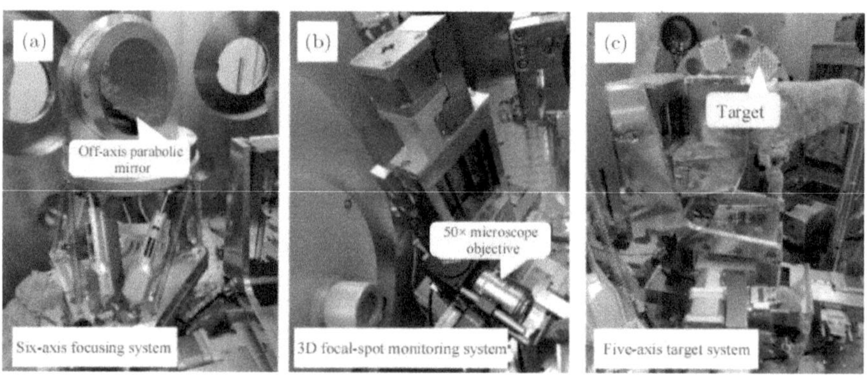

Fig. 8.8 (a) Six-dimensional focusing system; (b) three-dimensional focal spot monitoring system; (c) five-dimensional target system.

has high translational and rotational resolution, which can be used to correct the posture of the OAP to obtain the best focus. To achieve higher laser intensity, laser ion accelerators typically use short-focus OAPs with an f-number generally less than 3.

(2) Focal Spot Monitoring System

The three-dimensional focal spot monitoring system consists of a three-dimensionally movable translation adjustment system (x, y, z), a microscopic imaging system, and a CCD, as shown in Figure 8.8(b). The focal spot monitoring system uses a microscope objective with an appropriate magnification to enlarge and image the focal spot onto the CCD, which can meet the focusing adjustment requirements. In addition to optimizing the focus, the focal spot monitoring system can also be used to observe the target surface, achieving high-precision beam-target coupling. Figure 8.9 shows the laser focal spot and the target surface observed by the focal spot monitoring system.

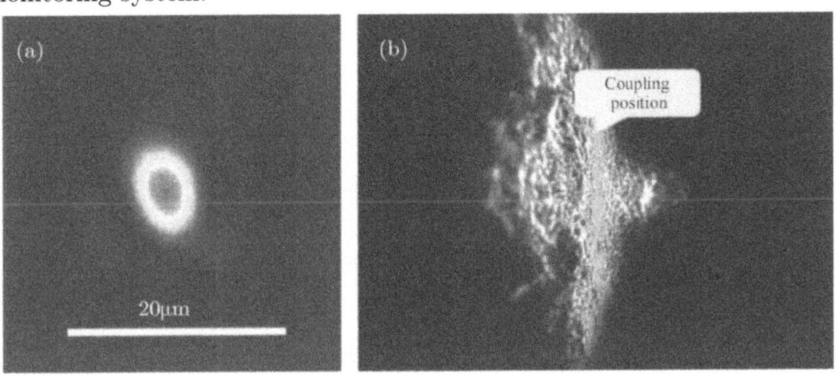

Fig. 8.9 (a) Laser focal spot; (b) target surface observed by the focal spot monitoring system.

(3) Target System

The five-dimensional target system is composed of a three-dimensionally movable translation adjustment system (x, y, z), a two-dimensional rotation system (θ_x, θ_y), and a target holder system, as shown in Figure 8.8(c). The three-dimensional translation system and the two-dimensional rotation system can adjust the target in five dimensions, and the target holder system is used to install the target surface. Different target holders can be designed according to specific target requirements. Moreover, the precision of beam-target coupling is also a major factor affecting the experimental results. Before each round of experiments, the focal spot monitoring system is used to correct each target surface to ensure that the target surface is at the focal point of the laser, which can achieve good beam-target coupling and provide a basis for effective ion acceleration and the repeatability of the experiment.

(4) Vacuum System

When high-power lasers are focused, nonlinear effects in the atmosphere dissipate the laser energy. Therefore, the ion acceleration system needs to operate in a vacuum environment. The vacuum level of the vacuum system generally needs to be better than 10^{-3} Pa. If high-vacuum diagnostic devices are required, the vacuum needs can be considered according to the actual situation.

(5) Plasma Parameter Diagnostic System

To better understand the process of laser ion acceleration, plasma parameters such as plasma density and plasma expansion velocity are usually diagnosed in experiments. Since the time scale of laser acceleration is very short, optical methods are commonly used to diagnose plasma parameters.

(6) Ion Diagnostic System

The ion diagnostic system is mainly used to diagnose the types of ions, energy, and ion numbers obtained from laser acceleration and is an important module in the ion acceleration experimental system. It will be discussed in detail in Section 8.3.

8.3 Introduction to Ion Diagnostic Systems

The ion diagnostic system is used to detect the types of ions, energy, and ion numbers obtained from laser acceleration, obtaining ion energy spectra. Ion energy spectrum diagnostics are very important for the study of ion acceleration mechanisms, so in the past two to three decades of laser ion acceleration research, its measurement methods have also developed rapidly. The ion diagnostic system includes ion detectors and detection media. The following introduces some common detection media and detectors.

8.3.1 *Detection Media*

Currently, the mature detection media mainly consist of: CR-39, radiographic film (RCF), imaging plate (IP), microchannel plate (MCP), and scintillator. The following introduces different detection media.

(1) CR-39

CR-39 is a commonly used ion detector, with a simple and lightweight structure, low cost, sensitivity to charged particles, low environmental stability requirements, and the ability to detect high-energy particles. It is currently widely used in research fields such as high-energy fission fragment measurement, fusion plasma diagnostics, cosmic rays, and neutron detection. Its chemical formula is $C_{12}H_{18}O_7$, a transparent material with a density of approximately 1.3 g/cm^3. The biggest advantage of CR-39 is that it does not respond to X-rays, hot electrons, and other products generated in laser acceleration experiments, thus having a unique advantage in ion diagnostics. When charged particles pass through CR-39, they leave atomic-scale radiation damage along their tracks. Then, the tracks are etched with strong acid or alkali solutions (usually NaOH or KOH solutions). The irradiated damage areas, having higher chemical activity, will be etched quickly, while the non-irradiated areas will be etched more slowly. After sufficient etching time, the tracks can be seen under a microscope. The main disadvantage of CR-39 in application is that it takes a long time to obtain experimental results, and it easily saturates in cases of high ion yields. Therefore, CR-39 is mainly suitable for low-ion-yield situations.

(2) Radiographic Film (RCF)

Radiographic film (Radiochromic Film, RCF) is a dosimeter widely used in medical physics, initially for high-energy photon dose detection and later applied to the field of laser ion acceleration. When ions are incident on RCF and deposit energy in it, RCF undergoes chemical changes, and its color changes. RCF is insensitive to the visible light region but is sensitive to various ionizing radiation sources. After irradiation, RCF turns dark blue. The absorbed dose of RCF is almost linear with its optical density over several orders of magnitude, so the absorbed dose and related charge information can be obtained by measuring the change in optical density before and after irradiation. Compared with CR-39, RCF has a relatively complex structure, generally consisting of a protective layer, a sensitive layer, and a base layer. Different models of RCF have different material thicknesses and structures, and their responses to radiation also vary, so calibration of optical density and absorbed dose is required according to the purchased RCF model.

Radiation Chromic Film (RCF) undergoes a noticeable color change after being irradiated, as depicted in Figure 8.10. Additionally, since RCF is sensitive to high-energy rays and electrons, a protective layer must be added during proton diagnostics to eliminate the influence of other products. RCF requires a certain dose of irradiation to initiate color change, hence its sensitivity is lower than that of CR-39.

(3) Imaging Plate (IP)

Imaging Plates (IP) were initially employed in medical X-ray diagnostics to convert analog image signals into digital ones. Similar to RCF, IPs are sensitive to protons, ions, electrons, X-rays, and gamma rays. Taking

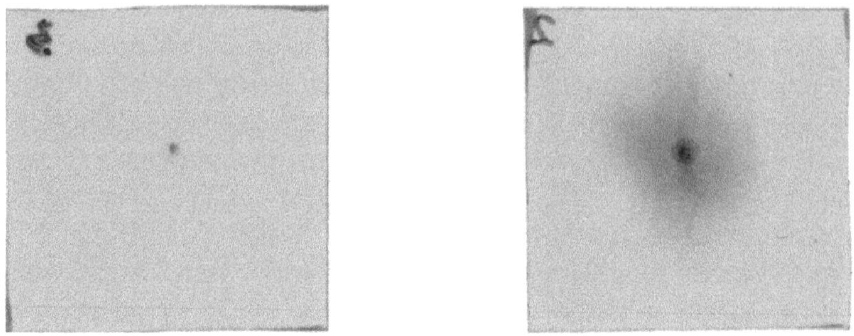

Fig. 8.10 RCF before and after irradiation (left: unirradiated, right: irradiated).

protons as an example, when protons strike the IP, they generate trapped electrons and holes on its surface, proportional to the proton beam dose at that point. The proton signal is stored in the form of electron-hole pairs. When scanned by an IP scanner, the trapped electrons and holes are released under the excitation of the scanning laser, producing fluorescence proportional to the density of electron-hole pairs at that point. The collected fluorescence is then converted into an electrical signal, amplified, and digitized for computer processing to generate a digital image.

IPs also have a multi-layer structure, including a protective layer, a fluorescent layer, a support layer, and a backing layer. The primary working medium is the fluorescent layer, typically composed of phosphorescent materials with high dynamic range and sensitivity. However, due to the short lifetime of the luminescent center energy levels, IPs need to be read promptly and corrected for decay rates. Moreover, different proton energies elicit varying responses from IPs, necessitating calibration. Figure 8.11 illustrates the measurement of protons using an IP, with the black line representing the response signal of protons deflected by a magnet and deposited on the IP.

(4) Microchannel Plate (MCP)

The three proton detection media mentioned earlier can only perform single-shot measurements in laser proton acceleration experiments and are not suitable for high-repetition-rate experiments. Here, we introduce a detection medium that can be used for online measurements-Microchannel Plate (MCP). The Microchannel Plate (MCP) is a multi-channel electron multiplier that evolved from the single-channel electron multiplier in the 1970s. Its working principle is similar to that of the single-channel electron multiplier. When a high voltage is applied across the channels, it forms

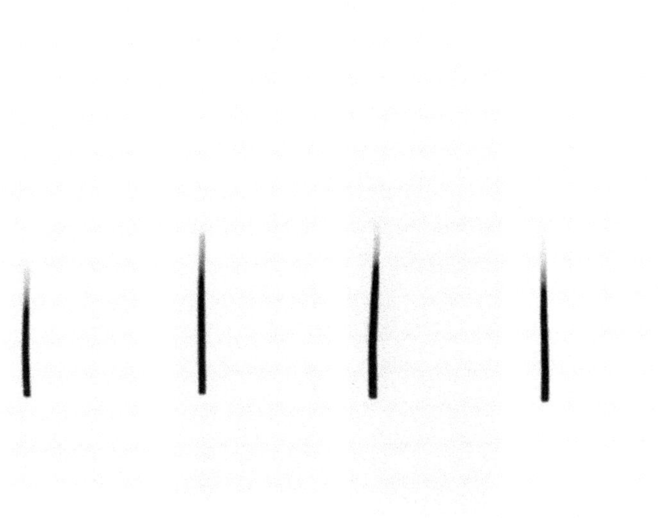

Fig. 8.11 Proton signal measured by IP in an experiment.

a continuous multiplication of the incident signal. When protons enter the channel, they collide with the inner surface of the channel, releasing secondary electrons. As these secondary electrons travel within the channel, they collide with the inner surface again, generating a large number of secondary electrons. This process repeats continuously within the channel, ultimately forming a multiplied electron beam at the channel output. The gain of the MCP mainly depends on the ratio of the channel length l_{mcp} to the channel diameter d_{mcp}, $\alpha_{\mathrm{mcp}} = l_{\mathrm{mcp}}/d_{\mathrm{mcp}}$, and the voltage applied across the channel ends. Typically, the gain of an MCP is on the order of 10^4 (depending on the applied voltage), as shown in Figure 8.12. For online measurement, a fluorescent screen must be added after the MCP to convert the electron signal into an optical signal. The visible light emitted by the fluorescent screen is then imaged onto a CCD through a lens to achieve online measurement. The disadvantages of MCPs include their relatively high cost, susceptibility to electromagnetic signals, and requirement for ultra-high vacuum. To prevent high-voltage breakdown of the MCP, a vacuum level below 1×10^{-4} Pa is generally required.

MCPs exhibit different responses to various ion species and incident angles, necessitating calibration in experiments. A calibration method using

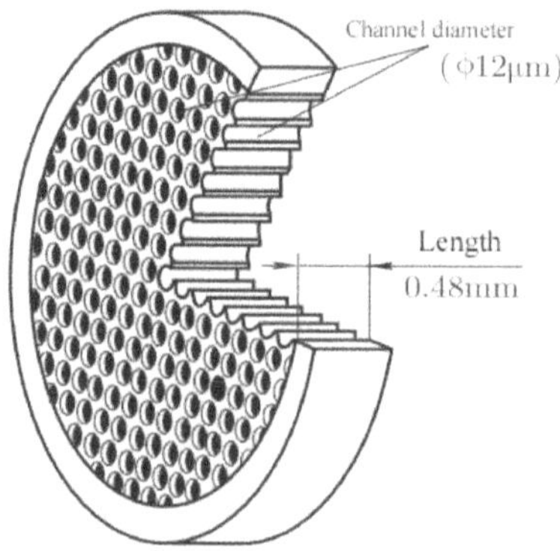

Fig. 8.12 Structure of a microchannel plate (MCP) [8].

^{241}Am to generate 5.48 MeV α particles has been reported [9], but this method is only applicable to a single energy. Additionally, CR-39 can be used for calibration by correlating the ion counts recorded on CR-39 with the MCP [10, 11].

(5) Scintillator

Similar to MCPs, scintillators can also be used for online measurements. They are organic plastics that respond to ions, electrons, neutrons, high-energy X-rays, and gamma rays. When particles strike the scintillator, secondary electrons are excited within it, leading to excitation or ionization. During the de-excitation process, the scintillator emits light instantaneously, with a rise time of approximately 0.5 ns and a total emission duration of about 1.5 ns [12]. This makes scintillators suitable for real-time measurement of ion or neutron spectra, as well as time-of-flight measurements [13]. Scintillators offer a wide measurement range, stable performance, durability, and are easy to use with minimal requirements for experimental conditions. In addition to time-of-flight measurements, they can be combined with Thomson spectrometers to measure ion energy spectra in laser ion acceleration experiments in real-time. Scintillators have widespread applications in medical physics, nuclear physics research, and astrophysics. Quantitative studies have shown that although scintillators have lower sensitivity than CR-39 and MCPs (for example, the detection threshold for 1.9 MeV protons is approximately 13×10^2 particles/(50 μm)^2, which is one order of magnitude lower than that of IP plates), they have a detection range spanning nearly four orders of magnitude and can respond linearly. This makes them more suitable for environments with relatively high ion radiation intensity. On the other hand, in terms of measurement precision, scintillators have higher resolution than most instruments used in radiation measurements, meaning their use will not degrade the overall resolution of the measurement system. The combination of scintillators with CCDs for real-time measurement is simple and feasible, and is expected to find widespread application in laser ion measurements.

In laser ion acceleration experiments, it is necessary to diagnose various properties of ion beams. Typically, multiple detection media are used in combination to achieve the best detection results.

8.3.2 *Detection Equipment*

(1) Thomson Spectrometer

The Thomson Parabola Spectrometer (TPS), proposed by Thomson nearly a century ago, was initially used for mass spectrometry measurements and has now becomc an important diagnostic device for laser-driven ion acceleration [14]. The greatest advantage of this spectrometer is that it can obtain the energy spectrum distribution of all ion species in a single measurement. The principle of the Thomson spectrometer is shown in Figure 8.13(a). When ions enter the spectrometer, they are subjected to both electric and magnetic forces. Since ions have different incident velocities and charge-to-mass ratios, the electromagnetic forces experienced by ions of different types and energies vary, resulting in different exit positions for the ions. The positions of the ions can be used to determine their types and energies. On the detection screen of the Thomson spectrometer, a large number of ions with different energies and charge-to-mass ratios form parabolas, as shown in Figure 8.13(b).

Ions in the spectrometer are subjected to both electric and magnetic forces, which can be described as

$$\boldsymbol{F} = q_{\mathrm{i}}(\boldsymbol{E} + \boldsymbol{v} \times \boldsymbol{B}), \tag{8.3.1}$$

where $\boldsymbol{E}$ is the electric field strength, q_{i} is the ion charge, $\boldsymbol{v}$ is the incident ion velocity, and $\boldsymbol{B}$ is the magnetic induction strength of the spectrometer. As shown in Figure 8.13(a), initially, ions move along the z-direction, with the electric and magnetic fields in the x-direction. Assuming the electric and magnetic fields are uniform and have a length L, and the length of the drift section is l_{f}, with the charged ion mass being m_{i}, the displacements

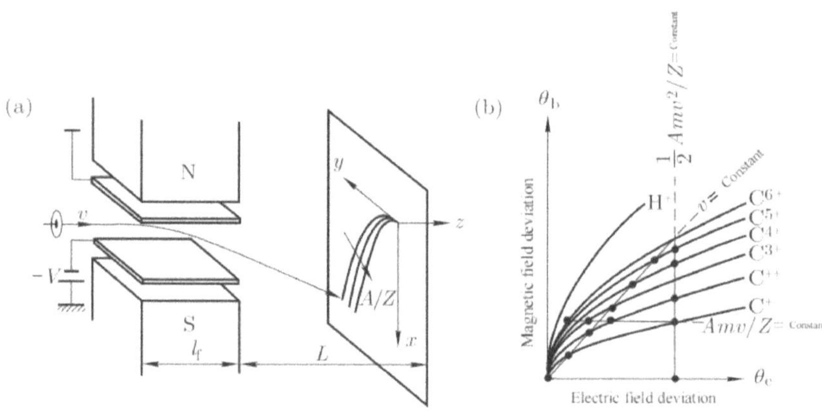

Fig. 8.13 (a) ; (b) .

of the charged ion in the x- and y-directions are

$$x = \frac{q_i EL}{m_i v^2}\left(\frac{L}{2} + l_f\right), \tag{8.3.2}$$

$$y = \frac{q_i BL}{m_i v}\left(\frac{L}{2} + l_f\right). \tag{8.3.3}$$

Combining the above two equations yields the expression for the parabola:

$$y^2 = \frac{q_i B^2 L}{m_i E}\left(\frac{L}{2} + l_f\right) x. \tag{8.3.4}$$

For ions at a specific position $(x,\ y)$, the ion's kinetic energy E_{kin} can be derived as

$$E_{\text{kin}} = \frac{[q_i BL(L/2 + l_f)^2]}{2m_i}\frac{1}{y^2}. \tag{8.3.5}$$

When the detector has a spatial interval of $[y,\ y+\Delta y]$, the corresponding energy interval is $[E_{\text{kin}},\ E_{\text{kin}} + \Delta E_{\text{kin}}]$. Since $\mathrm{d}N/\mathrm{d}E_{\text{kin}} = \Delta N/\Delta E_{\text{kin}}$, the ion energy spectrum can be obtained by calculating the number ΔN within the interval ΔE_{kin}.

(2) Angular-Resolved Spectrometer

While the Thomson spectrometer can distinguish the quantities of different types of particles based on their charge-to-mass ratios, it cannot obtain the angular distribution of proton beams. On the other hand, stacked detectors (see below) can provide the angular distribution of ion beams but cannot achieve online measurement. The angular-resolved spectrometer can realize the online measurement of particle angular distribution.

Refer to Figure 8.14. The angular-resolved spectrometer is similar to the Thomson spectrometer, with two differences: (i) the angular-resolved spectrometer has only a magnetic field and no electric field; (ii) the entrance aperture is replaced by a slit parallel to the magnetic field. By calculating the relative position of specific ions on the detection medium, the energy of ions, such as protons, landing at a particular location can be obtained.

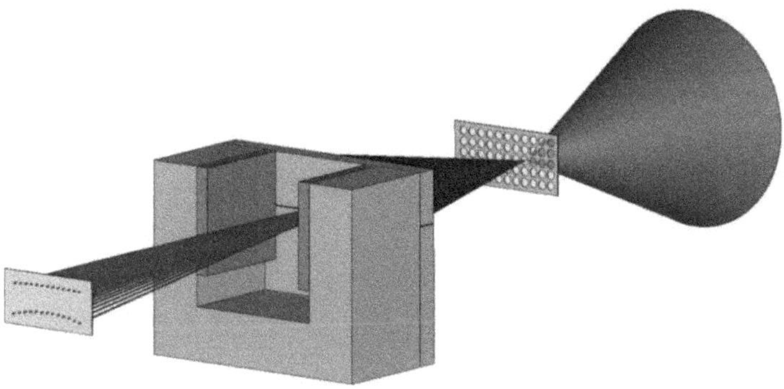

Fig. 8.14 Application of angular-resolved spectrometer in laser acceleration experiments [15].

(3) Stacked Detector

Although the Thomson spectrometer can obtain single-shot ion energy spectra online, its acceptance angle is typically on the order of μsr, and only a small fraction of ions are received by the spectrometer. Therefore, parameters such as the spatial distribution of ions, the conversion efficiency from laser to ions, the size of the ion source, divergence angle, and deflection angle cannot be directly obtained using the Thomson spectrometer. Thus, a stacked detector containing RCF and CR-39 is generally used to obtain these parameters. The stacked detector mainly acquires ion beam information through the energy deposition of ions in different layers. Since the energy loss process of ions exhibits a Bragg peak, the spatial and energy distributions of ions with different energies can be obtained in a single measurement. Furthermore, a deconvolution algorithm can be used to analyze the energy deposition in different layers of RCF to obtain ion energy spectrum information. The specific inversion algorithm can be found in the literature [7]. The SRIM [16] program is commonly used to calculate the ion deposition in different stacked configurations, thereby designing an appropriate stacking method. In experiments, the Thomson spectrometer is often used in conjunction with the stacked detector. Figure 8.15(a) shows a commonly used stacking method, while Figures 8.15(b)–(d) depict different models of RCF that can meet various dose measurement requirements.

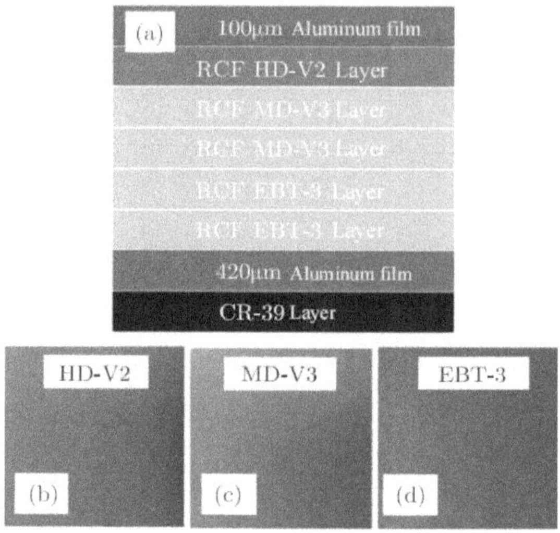

Fig. 8.15 (a) Stacking method of the stacked detector; (b)–(d): different models of RCF.

8.4 Laser Ion Acceleration Experiments

The previous sections have provided a brief introduction to the basic components of a laser ion accelerator. The following section will introduce typical laser ion acceleration experiments using the laser ion accelerator at Peking University as an example.

The experimental setup is shown in Figure 8.16(a). The laser energy on target in the experiment was approximately 1.6 J, with a pulse width of 30 fs. After optimization using a deformable mirror and a wavefront sensor, an $f/3.5$ off-axis parabolic mirror (OAP) was used to focus the laser, achieving a spot size of approximately 4.5 μm $\times$ 5.3 μm, with an energy concentration of 25%. The corresponding laser focal intensity was 8.3×10^{19} W/cm^2, and the normalized intensity a_0 was 6.2. The target material used was an aluminum foil with a thickness ranging from 0.8 to 6 μm.

Thomson spectrometers and stacked detectors are typically used to

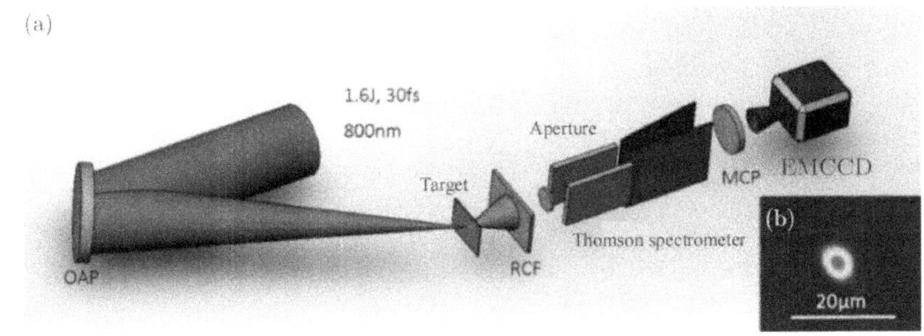

Fig. 8.16 (a) Experimental setup with the laser incident at 30° to the target; (b) laser focal spot (FWHM) of 4.5 μm × 5.3 μm [17].

diagnose the maximum proton energy, proton energy spectrum, and proton beam spatial distribution. After ions pass through the spectrometer, a microchannel plate (MCP) is used as the detector screen. The screen is imaged through a lens, and the signal from the detector screen is read using an electron-multiplying CCD (EMCCD) with 16-bit resolution to obtain experimental results online. A 30 μm aluminum foil is attached in front of the MCP to block transmitted laser light and low-energy carbon ions. The spectrometer is placed 140 mm away from the target, with an acceptance angle of 1.6×10^{-6} sr. The stacked detector is positioned 40 mm away from the target. To ensure detection of proton signals with different energies, three different models of RCF with varying sensitivities (HD-V2, MD-V3, EBT-3) are used.

Figure 8.17 shows typical experimental results obtained with a metallic target, specifically a 2.5 μm thick aluminum target. Figures 8.17(a)–(c) present the results from the RCF stacked measurements, with the white dashed line indicating the normal direction of the target. Using the SRIM program [16] to calculate the energy deposition of protons in different layers of the stacked detector, the main deposited proton energies on the three RCF layers are found to be 4.6 MeV, 5.9 MeV, and 7.9 MeV, respectively. There is still a significant proton signal on the 7.9 MeV RCF layer, corresponding to a proton beam charge of 5 pC. Figure 8.17(d) displays the proton energy spectrum measured by the Thomson spectrometer and RCF, with the curve representing the spectrum detected by the Thomson spectrometer. The cut-off energy of the proton beam has reached 10.2 MeV. The scatter plot represents the number of specific protons obtained based on the RCF calibration results and the reported spectral deconvolution

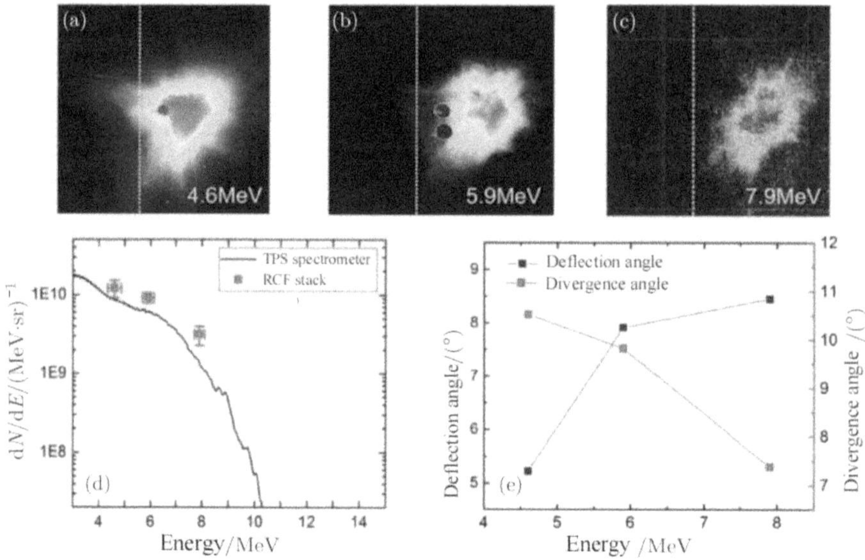

Fig. 8.17 Best acceleration results obtained with a metallic target in the experiment.

method in the literature [7]. Figure 8.17(e) illustrates the divergence and deflection angles of proton beams at different energies calculated from the proton beam spatial distributions in Figures 8.17(a)–(c).

References

[1] Geng Yixing. Experimental Research on Ultra-Short Ultra-Intense Laser Ion Acceleration [D]. Beijing: Peking University, 2017.

[2] Choi I. W., Jeon C., Lee S. G., *et al.* Highly efficient double plasma mirror producing ultrahigh-contrast multi-petawatt laser pulses [J]. *Opt. Lett.*, 2020, 45(23): 6342–6345.

[3] Rödel C., Heyer M., Behmke M., *et al.* High repetition rate plasma mirror for temporal contrast enhancement of terawatt femtosecond laser pulses by three orders of magnitude [J]. *Applied Physics B*, 2011, 103(2): 295–302.

[4] Nomura Y., Veisz L., Schmid K., *et al.* Time-resolved reflectivity measurements on a plasma mirror with few-cycle laser pulses [J]. *New Journal of Physics*, 2007, 9(1): 9–9.

[5] Tang Y., Hooker C. J., Chekhlov O. V., *et al.* Novel contrast enhancement of Astra Gemini laser facility [C]. CLEO: 2013. IEEE, 2013: 1–2.

[6] Wu Bo, Dong Ke Gong, Wu Yu Chi, *et al.* Identification of laser accelerated ion products using CR-39 tracks [J]. *High Power Laser and Particle Beams*, 2013, 25(2): 381.

[7] [7] Nürnberg F., Schollmeier M., Brambrink E., *et al.* Radiochromic film imaging spectroscopy of laser-accelerated proton beams [J]. *Review of Scientific Instruments*, 2009, 80(3): 033301.

[8] `http://www.hamamatsu.com.cn/product/17103.html` [Z].

[9] Sakabe S., Mochizuki T., Yamanaka T., *et al.* Modified Thomson parabola ion spectrometer of wide dynamic range [J]. *Review of Scientific Instruments*, 1980, 51(10): 1314–1315.

[10] Prasad R., Abicht F., Borghesi M., *et al.* Thomson spectrometer-microchannel plate assembly calibration for MeV-range positive and negative ions, and neutral atoms [J]. *Review of Scientific Instruments*, 2013, 84(5): 053302.

[11] Prasad R., Doria D., Ter-Avetisyan S., *et al.* Calibration of Thomson parabola-MCP assembly for multi-MeV ion spectroscopy [J]. *Nuclear Instruments and Methods in Physics Research Section A: Accelerators, Spectrometers, Detectors and Associated Equipment*, 2010, 623(2): 712–715.

[12] Flacco A., Sylla F., Veltcheva M., *et al.* Dependence on pulse duration and foil thickness in high-contrast-laser proton acceleration [J]. *Physical Review E*, 2010, 81(3): 036405.

[13] Ogura K., Nishiuchi M., Pirozhkov A. S., *et al.* Proton acceleration to 40 MeV using a high intensity, high contrast optical parametric chirped-pulse amplification/Ti: sapphire hybrid laser system [J]. *Opt. Lett.*, 2012, 37(14): 2868–2870.

[14] Li Chengyu. Design of Online Thomson Ion Spectrometer and Its Application in Laser Ion Acceleration [D]. Beijing: Peking University, 2014.

[15] Florian L. Online diagnostics for laser accelerated ions and electrons [D]. Munchen: Ludwig-Maximilians-University at Munchen, 2015.

[16] Ziegler J. F., Biersack J. P. The stopping and range of ions in matter [M]. Boston, MA: Springer, 1985: 93–129.

[17] Geng Y. X., Liao Q., Shou Y. R., *et al.* Generating proton beams exceeding 10 MeV using high contrast 60 TW laser [J]. *Chinese Physics Letters*, 2018, 35(9): 092901.

[18] Platt B. C., Shack R. History and principles of Shack-Hartmann wavefront sensing [J]. *Journal of Refractive Surgery*, 2001, 17(5): S573–S577.

[19] Bon P., Maucort G., Wattellier B., *et al.* Quadriwave lateral shearing interferometry for quantitative phase microscopy of living cells [J]. *Optics Express*, 2009, 17(15): 13080–13094.

Chapter 9

Introduction to Beam Transport Systems

Ion beams generated by laser acceleration typically exhibit large energy spread (100%), large divergence angles ($\sim 30^\circ$), and high peak currents (on the ampere scale). While significant theoretical and experimental progress has been made in laser ion acceleration, and new mechanisms to enhance ion energy and improve monoenergetic characteristics have been proposed, the energy and quality of ion beams directly produced by laser acceleration in current experiments still do not meet the demands of applications such as tumor therapy. Controllably regulating and transporting the ion beams is not only crucial for addressing issues related to energy spread, stability, and reliability of laser-accelerated beams but also represents an essential step in transitioning from "laser acceleration" to a "laser accelerator."

In the design of beam transport systems for laser-accelerated ion beams, it is essential to first incorporate focusing elements to efficiently capture the large-divergence ion beams and achieve a certain degree of beam collimation. Additionally, energy analysis components are necessary to obtain monoenergetic ion beams. Although the theory and application of beam transport systems for traditional accelerators are well-established for monoenergetic particle beams, they face new challenges when applied to the transport of laser-driven ion beams with large energy spread and large divergence angles.

This chapter will first introduce the fundamental knowledge of beam transport used in laser acceleration systems and some reported designs for laser ion beam transport. It will then delve into the design concepts and methods of laser ion beam transport systems, using the design of the beam transport system for the Laser Proton Accelerator at Peking University (CLAPA) and the results of beam experiments as examples.

9.1 Fundamentals of Beam Physics

Beam physics is the discipline that describes the morphology and motion laws of charged particle beams. The complete scope of beam physics research includes the morphology and motion laws of beams, the interaction and energy conversion between beams and electromagnetic waves, the interaction between beams and matter, the interactions among particles within a beam, the influences produced by beams on each other through the environment, and the changes in the state of particles within a beam, among other aspects. The beam transport in laser acceleration primarily focuses on the beam optics part: studying how to use electromagnetic fields to constrain and control the motion of charged particle beams, enabling particles to be transported along a designed trajectory. The emphasis is on constraining the particle trajectory, such as controlling the deflection, focusing, divergence, emittance growth, and imaging of particle beams. The research process of beam optics has certain similarities with optical transmission but is more complex compared to traditional optics.

In traditional accelerators, the longitudinal motion of the beam involves various aspects that need to be studied, such as energy changes, phase changes, and bunch stability. For laser accelerators, although there is a significant longitudinal bunching process in the radiation pressure acceleration, the longitudinal motion changes of the ion beam in the subsequent beam transport system are relatively simple. Therefore, for the beam transport system of a laser accelerator, the longitudinal motion is usually not the focus of research, but rather the transverse motion of the beam is the primary research issue.

9.1.1 *Phase Space and Emittance*

In beam optics, charged particle beams are typically studied as a whole. To better describe the motion state of charged particles, the concept of phase space must first be introduced. For a charged particle moving in a three-dimensional Cartesian coordinate system $Oxyz$, once the three position coordinates x, y, z and the three momentum components p_x, p_y, p_z of a particle are determined, the motion state of the particle will be completely defined. The six-dimensional space composed of this set of three-dimensional position coordinates and three dimensional momentum coordinates is called phase space. Any motion state of a particle can be represented by a point in the six-dimensional phase space.

When describing the motion state of particles in a certain cross-section at a longitudinal position $z = z_i$, the six-dimensional phase space can be simplified to a four-dimensional phase space composed of x, y, p_x, and p_y. For particle beams whose motion in the x and y directions is uncoupled, two independent two-dimensional phase planes can be used to describe them. For rotationally symmetric beams with circular cross-sections, the transverse motion state of particles can be described using a two-dimensional phase plane composed of r and p_r.

In the right-hand side of Figure 9.1, AB indicates the boundary of the beam cross-section at the longitudinal position z_i. The points on M_1M_2 represent the motion states of all particles at point M in the beam cross-section at $z = z_i$, which is a distance r_M from the beam axis; the points on N_1N_2 represent the motion states of all particles at point N, which is a distance r_N from the beam center axis. Therefore, the points within the ellipse in the phase diagram can represent the motion states of all particles in the beam cross-section at $z = z_i$. This ellipse is called the particle beam's emission phase diagram at $z = z_i$. The area enclosed by the ellipse is defined as the emittance phase space area (hereinafter referred to as the phase area). The transverse coordinates r_A and r_B determined by the ellipse boundary represent the radial size of the beam, while the maximum vertical coordinate of the ellipse boundary can represent the spatial solid angle of the beam. The beam emittance is usually defined as the ratio of the beam's phase space area A to π, and can be expressed as follows, with the commonly used units being m·rad, cm·rad, or mm·mrad:

$$\varepsilon = \frac{A}{\pi}. \tag{9.1.1}$$

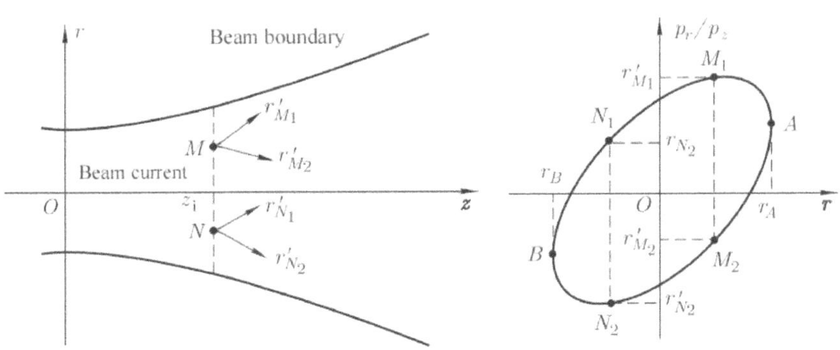

Fig. 9.1 Phase diagram of particle motion at the beam cross-section at $z = z_i$.

However, for actual beams, particles are generally not uniformly distributed, and it is difficult to define the exact position of the boundary. Therefore, in practical measurements and calculations, the root-mean-square emittance (ε_{rms}) is used, defined as follows:

$$\varepsilon_{\text{rms}} = \sqrt{\langle x^2\rangle\langle x'^2\rangle - \langle x \cdot x'\rangle^2}. \tag{9.1.2}$$

Considering the comparability of emittance for particle beams with different energies and removing the effect of beam velocity on emittance size, the normalized root-mean-square emittance ($\varepsilon_{\text{n,rms}}$) is defined as

$$\varepsilon_{\text{n,rms}} = \beta\gamma\sqrt{\langle x^2\rangle\langle x'^2\rangle - \langle xx'\rangle^2}, \tag{9.1.3}$$

where β is the normalized velocity of the particle, and γ is the relativistic factor.

The traditional definition of emittance is usually applicable to quasi-monoenergetic beams, where particle energy is uncoupled from angle. However, this is not the case for laser proton beams. For laser-driven ion beams, Migliorati *et al.* [1] proposed an extended form of the normalized root-mean-square emittance:

$$\varepsilon_{\text{n,rms}} = \sqrt{\langle x^2\rangle\langle(\beta\gamma x')^2\rangle - \langle x\beta\gamma x'\rangle^2}. \tag{9.1.4}$$

This formula takes into account the energy differences among different particles and is applicable to both traditional accelerators and laser accelerators. Considering the possibility of measurement errors in experimental measurements, Floettmann *et al.* [2] proposed two definitions of emittance. The first is the normalized phase space (x, p_x) root-mean-square emittance:

$$\begin{aligned}\varepsilon_{\text{n,rms}} &= \sqrt{\langle x^2\rangle\langle(\beta\gamma x')^2\rangle - \langle x\beta\gamma x'\rangle^2} \\ &= \frac{1}{m_0 c}\sqrt{\langle x^2\rangle\langle p_x^2\rangle - \langle xp_x\rangle^2},\end{aligned} \tag{9.1.5}$$

where m_0 is the particle mass. This definition of emittance is equal to the extended emittance formula proposed by Migliorati *et al.* when the ion beam is forward-propagating ($p \approx p_z$). The second is the normalized trace space (xx') emittance:

$$\begin{aligned}\varepsilon_{\text{n,tr,rms}} &= \frac{\overline{p}_z}{m_0 c}\sqrt{\langle x^2\rangle\langle x'^2\rangle - \langle xx'\rangle^2} \\ &= \frac{\overline{p}_z}{m_0 c}\sqrt{\langle x^2\rangle\left\langle\left(\frac{p_x}{p_z}\right)^2\right\rangle - \left\langle x\frac{p_x}{p_z}\right\rangle^2}.\end{aligned} \tag{9.1.6}$$

It can be seen that after the inclusion of the energy spread coupling term, particle beams with different energies disperse due to phase space rotation, leading to emittance growth. The normalized phase space root-mean-square emittance is clearly no longer a conserved quantity here; whereas the normalized trace space emittance is always conserved in drift space. It should be pointed out that the normalized trace space emittance does not have a clear physical meaning.

According to the definition of normalized phase space root-mean-square emittance, ignoring the coupling between energy and transverse angle, the following derivation can be made:

$$\begin{aligned}\varepsilon &= \langle x^2\rangle\langle \beta^2\gamma^2 x'^2\rangle - \langle x\beta\gamma x'\rangle^2 \\ &= \langle \beta^2\gamma^2\rangle\langle x^2\rangle\langle x'^2\rangle - \langle\beta\gamma\rangle^2\langle xx'\rangle^2 \\ &= (\langle\beta^2\gamma^2\rangle - \langle\beta\gamma\rangle^2)\langle x^2\rangle\langle x'^2\rangle + \langle\beta\gamma\rangle^2(\langle x^2\rangle\langle x'^2\rangle - \langle xx'\rangle^2).\end{aligned} \tag{9.1.7}$$

Considering the initial emittance definition

$$\varepsilon_0^2 = \langle x^2\rangle\langle x'^2\rangle - \langle xx'\rangle^2, \tag{9.1.8}$$

finally,

$$\varepsilon = (\beta^2\gamma^2 - \langle\beta\gamma\rangle^2)\langle x^2\rangle\langle x'^2\rangle + \langle\beta\gamma\rangle^2\varepsilon_0^2, \tag{9.1.9}$$

where the second term on the right-hand side corresponds to the normalized root-mean-square emittance in conventional accelerators, which is a conserved quantity; while the first term on the right-hand side is an additional term unique to laser accelerator beams. For beams with large energy spread, it can be clearly seen that the emittance is not conserved. This emittance growth due to energy spread and divergence angle has been verified in simulations [3] and observed in experiments measuring the emittance of laser proton beams [4]. Similar emittance growth has also been reported in studies of laser-driven electron beams [37].

9.1.2 *Transverse Motion of the Beam*

In laser accelerator facilities, laser-accelerated beams exhibit large divergence angles and large energy spreads, with a certain coupling between energy and divergence angle. This makes the stability of transverse motion particularly important in beam transport systems. Typically, transverse focusing in laser accelerator beam transport systems is achieved through magnetic and electric fields, with electric field focusing being more suitable for low-energy ion beams. Existing laser accelerator beam transport systems mainly rely on magnetic field focusing.

Based on the principles of traditional accelerator physics, the orbit equation for the transverse motion of particles is as follows [5,6]:

$$\begin{cases} \dfrac{\mathrm{d}^2 x}{\mathrm{d}s^2} + (1-n)\dfrac{x}{\rho^2} = 0, \\ \dfrac{\mathrm{d}^2 z}{\mathrm{d}s^2} + n\dfrac{x}{\rho^2} = 0, \\ n = -\dfrac{r}{B}\dfrac{\partial B}{\partial r}, \end{cases} \tag{9.1.10}$$

where n is the magnetic field gradient coefficient, also known as the logarithmic gradient of the magnetic field, which is a parameter indicating the variation of the magnetic field with radius and can be expressed in terms of the magnetic field strength and the radius of the magnet. The solution to the above equations can be written as:

$$\begin{cases} x = \sqrt{x_{\mathrm{i}}^2 + \dfrac{\gamma_{x\mathrm{i}}^2 \rho_{\mathrm{i}}^2}{1-n}} \cos\left(\dfrac{\sqrt{1-n}}{\rho} s + \varphi_x\right) + x_{\mathrm{c}}, \\ z = \sqrt{z_{\mathrm{i}}^2 + \dfrac{\gamma_{z\mathrm{i}}^2 \rho_{\mathrm{i}}^2}{n}} \cos\left(\dfrac{\sqrt{n}}{\rho} s + \varphi_z\right), \end{cases} \tag{9.1.11}$$

where $(x_{\mathrm{i}}, \gamma_{x\mathrm{i}})$, $(z_{\mathrm{i}}, \gamma_{z\mathrm{i}})$, and ρ_{i} are the initial position, initial incidence angle, and initial orbit curvature radius of the charged particle entering the accelerator, respectively. ρ and φ_x, φ_z represent the orbit curvature radius and initial phase of the particle, and x_{c} is the orbit dispersion caused by the momentum dispersion Δp of the charged particle, which can be expressed as:

$$\begin{cases} x_{\mathrm{c}} \approx \dfrac{\rho}{1-n}\dfrac{\Delta p}{p}, \\ \rho = \dfrac{\sqrt{W(W+2\varepsilon_0)}}{cqB}, \end{cases} \tag{9.1.12}$$

where W, ε_0, and q are the total energy, rest energy, and charge of the particle, respectively. It can be seen that when the particle energy is very high, the orbit curvature radius ρ of the particle is large. This leads to a large amplitude of transverse particle motion, and the momentum dispersion term further increases the amplitude of transverse motion. If the ion beam cannot be effectively controlled, the beam will quickly be lost in the vacuum pipe.

Drawing inspiration from optical propagation: when a beam of light propagates through an optical lens system, concave and convex lenses play the roles of defocusing and focusing the light beam. By properly arranging and combining them, a very good focusing system for the entire transmission system can be achieved. If we analogize the particle beam to a light beam and the magnetic lenses with positive and negative magnetic field gradients to concave and convex lenses alternately arranged along the particle orbit, then the particle beam can also achieve good focusing transmission effects by properly arranging the magnetic lenses. This theory is known as the strong focusing principle of beams and is now widely used in various conventional radiofrequency accelerators and is also applicable to the design of laser-driven particle beam transport.

In the actual design process, focusing structures are usually chosen flexibly according to the characteristics of the beam itself and the different requirements of the beam transport system. Table 9.1 lists some common beam focusing elements [5]. In the conventional accelerator design process, it is also common to convert the transmission elements into matrix forms to represent the linear optical changes in the beam transport process. However, for laser accelerators, due to the large energy spread, the beam phase

Table 9.1 Comparison of common beam focusing elements.

Focusing Element	Focal Length	Illustration
Quadrupole lens	$f_x = \frac{1}{GL},\ f_y = \frac{-1}{GL}$	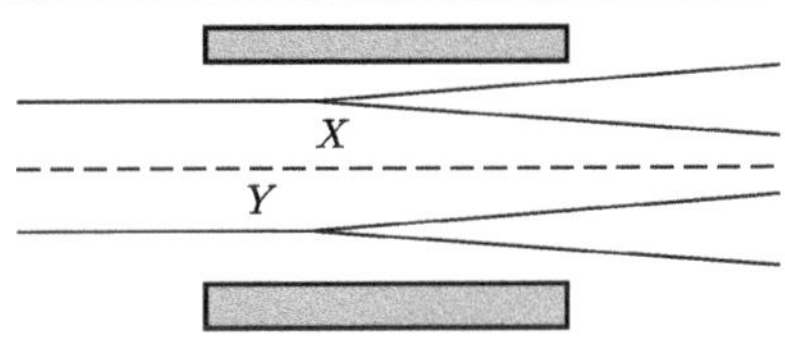
Electrostatic lens	$f = \dfrac{1}{\dfrac{1}{8\sqrt{V}}\displaystyle\int_{z_0}^{z_1}\frac{1}{V^{3/2}}\mathrm{d}z}$	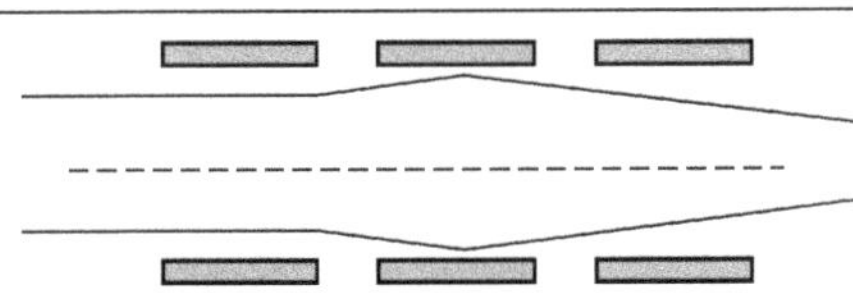
Solenoid	$f = \dfrac{1}{\dfrac{e^2}{4p^2}\displaystyle\int_{z_0}^{z_1}B(z)^2\mathrm{d}z}$	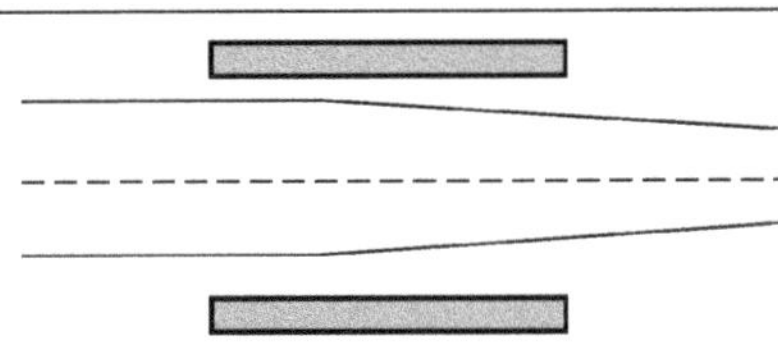

Note: G: Field gradient of quadrupole lens; L: Distance between quadrupole lens centers; V: Electric potential of electrostatic lens; p: Canonical momentum of ions; B: Field strength of solenoid.

space distortion is relatively severe, and higher-order transmission matrices need to be considered. The analysis method of beam transport matrices can only be used as a reference. This will not be discussed in depth here.

From Table 9.1, it can be seen that solenoids and electrostatic lenses are axially symmetric, with the same focusing effects in the X and Y directions. However, quadrupole lenses composed of quadrupole magnets can only focus in one direction and defocus in the other. Therefore, multiple quadrupole magnets need to be combined to achieve focusing in both directions.

9.1.3 *Beam Dispersion Function and Energy Spread Effects*

For particle beams produced by laser acceleration, an important physical quantity involved in the design of beam transport systems is the dispersion function. During the transport of a particle beam, a reference particle with a given energy, usually the average or central energy of the entire beam, is typically specified. Particles with energy deviations from the central energy will have different deflection angles when subjected to a magnetic field. When a beam with a certain energy distribution passes through a deflecting magnetic field, particles with energy deviations will experience an angular perturbation $\theta\Delta p/p_0$, projecting the longitudinal energy spread onto the transverse scale. This characteristic is commonly used in traditional accelerators for beam energy analysis and is also applied in the subsequent energy selection systems of laser accelerators.

Considering higher-order terms, the transverse motion Eq. (9.1.10) can also be written in the following form:

$$x'' - \frac{\rho + x}{\rho^2} = -\frac{B_y}{B\rho}\left(1 + \frac{x}{\rho}\right)^2. \tag{9.1.13}$$

For a non-ideal particle with momentum p, the motion equation can be written as

$$x'' - \frac{\rho + x}{\rho^2} = -\frac{B_y}{B\rho}\left(1 + \frac{x}{\rho}\right)^2 \frac{p_0}{p}. \tag{9.1.14}$$

Introducing a quadrupole field that varies linearly with transverse position, we have

$$B_y = B_0 + B'x. \tag{9.1.15}$$

Substituting this into the motion Eq. (9.1.14) for a non-central particle and neglecting higher-order terms of x/ρ, we obtain

$$x'' + \left(\frac{1}{\rho^2}\frac{2p_0 - p}{p} + \frac{B'}{B\rho}\frac{p_0}{p}\right)x = \frac{1}{\rho}\frac{\Delta p}{p}, \tag{9.1.16}$$

where $\Delta p \equiv p - p_0$. For a particle with momentum deviation, the change in the particle's orbit due to the momentum deviation Δp can be expressed as:

$$\begin{cases} x = x_0 + D_x \dfrac{\Delta p}{p}, \\ D_x'' + \left(\dfrac{1}{\rho^2} \dfrac{2p_0 - p}{p} + \dfrac{B'}{B\rho} \dfrac{p_0}{p} \right) D_x = \dfrac{1}{\rho}, \end{cases} \tag{9.1.17}$$

where D_x is the dispersion function, and x and x_0 are the distances of the particle from the central particle with and without dispersion, respectively. It can be seen that the dispersion acts on the beam, causing a coupling between the transverse and longitudinal directions of the beam. The momentum spread of the beam directly leads to an increase in the transverse envelope of the beam. For the beam as a whole, the change in the transverse envelope of the beam under dispersion can be expressed as

$$a_x = \sqrt{a_{x,0}^2 + \left(D_x \frac{\Delta p}{p} \right)^2}. \tag{9.1.18}$$

In the six-dimensional phase space, the coupling between the transverse and longitudinal directions does not increase the volume of the six-dimensional phase space, but it changes the emittance of the projected space, thereby leading to an increase in the beam envelope. If the beam still has dispersion after passing through the arc section, it will make the beam position very sensitive to the jitter of the central beam energy, significantly reducing the acceptance of the beam transport system. In traditional accelerators, beam transport line designs usually try to adopt dispersion-free designs. Laser accelerator beam transport systems, due to their inherently large momentum spread, are very sensitive to dispersion. However, most existing laser accelerator transport systems, due to their short transport distances and lack of very strict beam control and transport requirements, have mostly used simple transport units for design. From the expression of dispersion (9.1.17), it can be seen that the dispersion function changes similarly to the horizontal displacement of off-axis particles outside the deflecting magnetic field. Therefore, by arranging the magnets in the arc section according to certain rules to form a locally dispersion-free structure, that is, by limiting the dispersion function to appear only in a certain section of the beam transport system, or only in the deflecting section, and by controlling the dispersion function to be zero in most cases during the beam transport process, the impact of dispersion and energy spread on beam transport can be reduced.

9.1.4 *Barber's Rule*

Deflecting magnets are the most commonly used energy analysis elements. Their magnetic field strength, deflection radius, deflection angle, edge angle, and other factors all affect the deflection behavior of particles. To select particles within a certain energy range from a mixed beam of different energies or types of particles, it is first necessary to spatially separate particles of different energies. To improve the utilization efficiency of laser-driven particle beams, it is best to also gather particles of the same energy. Sector magnets have certain advantages in this regard and are very suitable for use in laser-accelerated ion beam transport systems. For sector magnets without edge angles, the behavior of particle beams complies with the Barber's Rule described below: If the waist of the particle beam in the x direction coincides with the object point, particles of the same energy can be gathered at the image point while particles of different energies are separated in the x direction. The object point and image point settings are the most important parameters for sector magnets.

Referring to Figure 9.2, Barber's Rule can be stated as follows: For a uniform magnetic field deflecting magnet with zero edge angle, in the radial plane, the object point of the magnet, the center of curvature of the reference orbit, and the image point must all lie on the same straight line. This requires that the object distance and image distance satisfy a certain geometric relationship. In layman's terms, if the particle source or beam waist is vertically incident at the object point of the deflecting magnet, particles of the same energy but different divergence angles will converge at the image point after deflection. Beyond the image point, particles of the same energy begin to diverge, leading to mixing with particles of other energies and resulting in inaccurate energy selection. After deflection, particles of different energies are easily separated. As long as particles of the same energy converge at the image point, relatively accurate energy selection can be achieved. At the image point, where different energy particles are

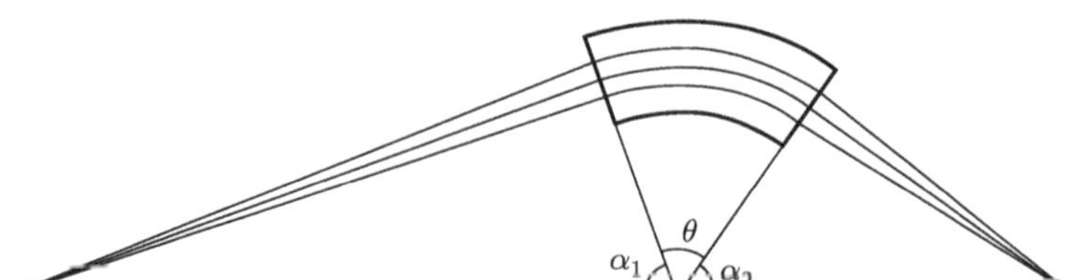

Fig. 9.2 Schematic diagram of Barber rule [38].

separated and particles of the same energy converge, a slit can be placed. By controlling the size of the slit opening, a specific range of particle energies can be selected. In practice, particles of different energies have momentum dispersion after passing through the deflecting magnet, resulting in different deflection radii and image point positions. Therefore, in actual design, the different image point positions of particles with different energies are usually not considered, and the image point position of the central energy particles is used as the standard to set the slit for energy selection, which is the commonly used one-dimensional slit. A one-dimensional slit is not very accurate for energy selection in beams with large energy spread, but it is more convenient in mechanical machining and actual operation.

9.2 Common Elements in Laser Accelerated Beam Transport Systems

9.2.1 *Beam Focusing Elements*

The ion beam bunches produced by laser acceleration have dimensions on the micrometer scale, but their divergence angles can be as high as several degrees to tens of degrees. Therefore, careful consideration must be given to the collection of large-angle ion beams. In existing designs of laser-accelerated beam transport systems, the most commonly used focusing elements are quadrupole lenses and solenoids. Quadrupole lenses composed of quadrupole magnets, either in binary or ternary configurations, have been widely used in the collection of laser-accelerated electrons or ions. Unlike quadrupole magnets, solenoids can focus particles in both transverse directions, which is advantageous for capturing highly divergent particle beams. Pulsed or superconducting solenoids have therefore been extensively studied [7–10].

Quadrupole magnets, based on the strong focusing principle, produce a strong focusing force in one direction and a defocusing force in the other. Two or three quadrupole magnets combined into a quadrupole lens can achieve focusing in both transverse directions under suitable magnetic field parameters. In 1998, Dewa used a quadrupole lens to focus high-energy electrons from laser wakefield acceleration [11], and since then, quadrupole lenses have been widely used for focusing and transporting laser-driven electron beams [12–14]. In 2008, Schollmeier *et al.* first reported the application of permanent magnet quadrupole lenses for focusing and transporting laser-accelerated ion beams [15]. As shown in Figure 9.3, the permanent

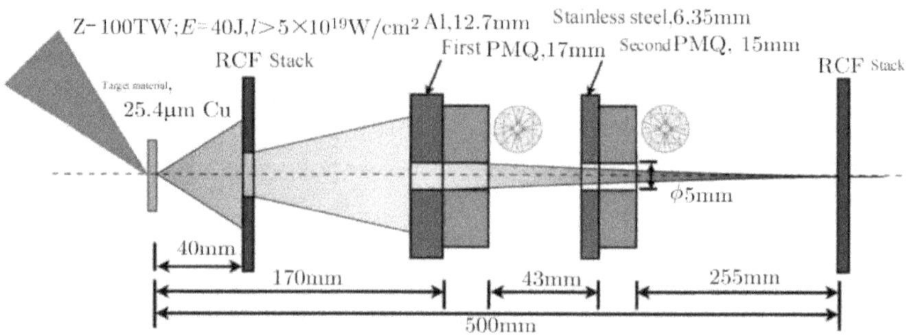

Fig. 9.3 Schematic diagram of a binary quadrupole lens collecting laser-accelerated ion beams [15].

magnet quadrupole lens (PMQ) has an aperture corresponding to a divergence angle of approximately ±15 mrad. A proton beam with an energy of (14 ± 1) MeV was collected by the quadrupole lens, and about 0.1% of the protons were detected on a radiochromic film (RCF) placed 50 cm from the target. The density of protons within this energy range increased by 75 times after focusing, indicating that although the collection efficiency of the quadrupole lens is not high, it still has strong focusing capabilities.

In 2009, Nishiuchi *et al.* [16] also conducted similar research using a binary quadrupole lens, as shown in Figure 9.4.

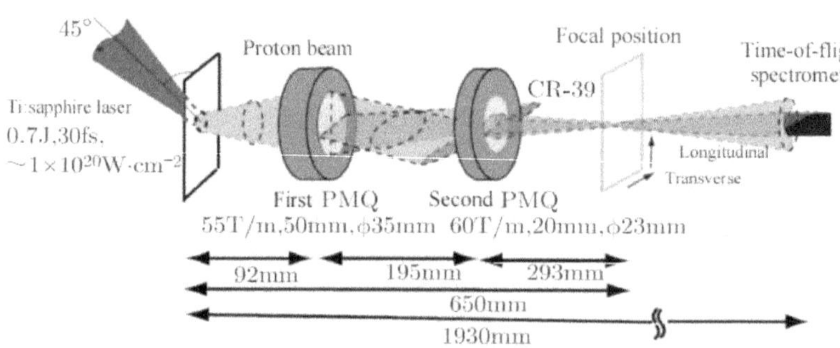

Fig. 9.4 Collection of laser-accelerated proton beams by a binary quadrupole lens, measurement of transverse proton distribution at different locations using CR-39 [16].

In this experiment, the transverse distribution of the proton beam was measured using CR-39 at distances of 450, 550, 650, 750, 850, and 927 mm from the laser target. The proton beam spot size is smallest at a distance of 650 mm from the laser target, where the beam waist is formed. Beyond this point, the proton beam begins to diverge.

It can be seen that quadrupole lenses have strong focusing power, but their symmetry in the transverse plane is not good. In particular, it is difficult to adjust the magnetic field strength of permanent magnet quadrupole lenses, resulting in a non-radially symmetric proton beam distribution. Solenoids have weaker focusing power but can achieve symmetric focusing in both transverse directions. For laser-accelerated ion beams with large divergence angles, the collected proton beam is symmetric in the transverse plane after focusing by a solenoid, making it suitable for use as a pre-injector for the next stage of acceleration in laser acceleration.

In 2009, Roth *et al.* [17] used a pulsed solenoid with a magnetic field strength of 8 T to collect laser-accelerated proton beams. Figure 9.5 shows the results of numerical simulations using the CST software. The solenoid was placed 1.7 cm behind the laser target to receive the particle beam produced by laser acceleration. A (5×5) cm radiochromic film (RCF) was placed 24 cm behind the laser target to detect the proton beam. Without solenoid focusing, only 23% of the protons were detected due to the large initial divergence angle. With the solenoid in use, approximately 81% of the particle beam could be collected. It is worth noting that some low-energy accompanying electrons produced during laser-plasma interactions were also collected and focused by the solenoid along with the proton beam. Since these low-energy electrons were closer to the central axis of the solenoid, they further enhanced the focusing capability of the solenoid.

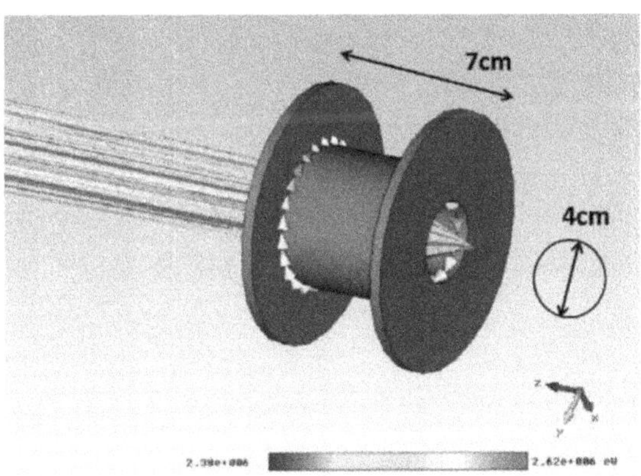

Fig. 9.5 CST simulates solotron collimating proton beam. The solenoid is 7 cm long and has an inner diameter of 4 cm. 8 T magnetic field strength collimating the 2.5 MeV proton beam [17].

Solenoids, in addition to focusing, can also perform some energy analysis through chromatic effects. In 2012, Hofmann *et al.* [18] reported a design using a solenoid in conjunction with a circular aperture slit. The slit was placed between the solenoid and the laser to control the divergence angle of the particle beam entering the solenoid, and another slit was placed after the solenoid to control the beam envelope. The energy spectrum and transverse ion density distribution of a proton beam with a central energy of 200 MeV and a divergence angle of 50 mrad after passing through the solenoid for focusing and energy analysis, and then through the slit. The ion density is higher closer to the center after solenoid focusing.

The advantage of the solenoid's symmetric focusing in the transverse plane is significant. Laser-accelerated ion beams, after being focused in the transverse direction, can pass through a radiofrequency (RF) cavity for energy spectrum compression, serving as a pre-injector for traditional accelerators. In 2013, Teng *et al.* [19] reported the use of a combination of a solenoid and an RF cavity to compress the energy spread of laser-driven proton beams. As shown in Figure 9.6, the experimental results indicate that after focusing by the solenoid, protons with a divergence angle less than 12° can be completely captured by the RF cavity. The electric field within the RF cavity can compress the beam spread around the central energy to an energy spread of 2.5% within a quasi-monoenergetic range.

The two types of focusing elements mentioned above, each with its own characteristics, can both be used for the collection of laser-accelerated

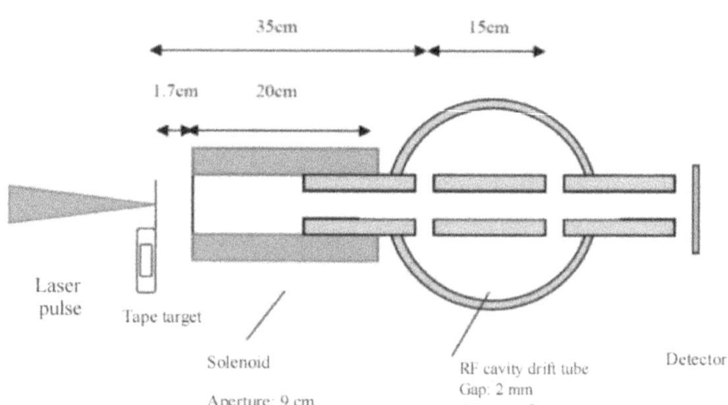

Fig. 9.6 After laser accelerated ion beam is collimated by a solenoid, and the beam energy dispersion near the center energy is compressed by a radio frequency (RF) cavity [19].

beams. With the in-depth research on laser-accelerated beam transport, and in combination with the actual engineering design of laser acceleration facilities, these two elements can be combined according to their respective characteristics for the design of beam transport systems.

9.2.2 *Beam Deflection Elements*

After the laser ion beam is collected and focused, it often needs to undergo energy analysis to select monoenergetic beam bunches that meet application requirements. The most commonly used energy selection element is the deflecting analysis magnet. The principle, as previously described, is to separate particles of different energies in space by utilizing the different deflection radii of charged particles in a magnetic field. After a certain angle of deflection, different energy particles are spatially separated. Subsequently, through slits or other devices, particles within a specific energy range can be selected and analyzed.

One common type is shown in Figure 9.7, consisting of four square dipole magnets with the middle two having the same magnetic field direction [20]. The first two dipole magnets are responsible for separating particles of different energies. A slit is placed in the middle to select particles within a certain energy range. The last two dipole magnets deflect the particles again, causing particles of different energies to converge back to a collimated state. The advantage of this device is that after energy selection, particles of different energies can still converge together, returning to an initially collimated state with similar beam spot sizes. The disadvantage is that when the initial particles have a divergence angle, it leads to mixing of particles of different energies at the selection point, affecting the analysis

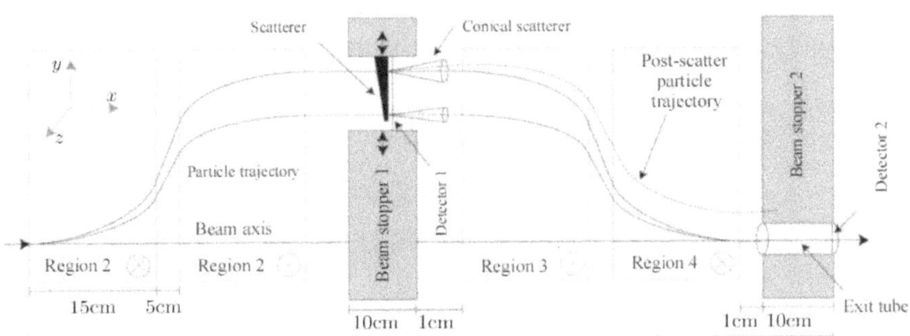

Fig. 9.7 Energy analysis system composed of four bipolar magnets [20].

accuracy. Therefore, this device is suitable for particle beams with good collimation, such as laser-accelerated electrons. If used for high-precision ion beam analysis, only a small part of the divergence angle can be selected, sacrificing collection and transmission efficiency.

After different energy protons are separated, not only is a selectable range of energy spectra obtained, but also the number and current of protons are changed, which is important for their applications in certain fields. For example, in the energy spectrum of laser-accelerated ion beams, the number of ions often increases with decreasing energy. However, the spread-out Bragg peak (SOBP) required in medical applications demands a higher number of ions in the high-energy region, and different depths of tumors require different ion energy spectrum ranges. Therefore, based on this device, Schell *et al.* [21] proposed placing energy absorbers of different thicknesses and scattering cones at the selection point to shape the energy spectrum. This method can expand the energy range through absorbers, but using scattering cones to reduce the number of low-energy particles lacks the predictability and operability for flexible adjustment of SOBP for different depths.

Another common energy analysis element is a sector magnet deflected at a certain angle, which is usually combined with quadrupole lenses [22] or solenoids to transport the beam. The sector magnet has a focusing effect in the radial direction, and the object and image distances are determined by Barber's Rule introduced in Section 9.1.4. After deflection, particles of the same energy but different scattering angles converge at the image point, while particles of different energies are spatially separated, allowing for energy analysis of beams with large divergence angles. In actual design, it is also necessary to select more complex beam manipulation elements such as beam uniformization systems according to actual engineering requirements to complete the final design of the beam transport system.

9.2.3 *Beam Diagnostic Devices*

Beam transport systems can achieve focusing and energy analysis of proton beams with large energy spread and large divergence angles, but they still lack diagnostics of the beam, such as whether the beam bunch deviates from the central trajectory, the amount of charge in the proton beam bunch, and the shape of the beam bunch. The basic layout principle of beam diagnostics in laser-accelerated transport systems is similar to that of traditional accelerators, but the actual setup is adjusted accordingly according to the needs. Existing beam transport systems, when considering the layout of beam diagnostics, basically start from the perspectives of beam position, beam current, and beam bunch shape. The following introduces some common beam diagnostic devices.

The integrating current transformer (ICT) is a device for measuring the charge of short beam pulses and is commonly used in particle accelerators and other high-energy physics experiments. Inside the ICT, there is a coil that generates an induced electromotive force when the beam passes through it, causing a change in the magnetic flux within the coil. The ICT also contains an integrating circuit composed of a capacitor and an amplifier that can convert the generated voltage signal into a charge signal, as shown in Figure 9.8. The measured charge is proportional to the initial beam charge, thus obtaining the beam's charge.

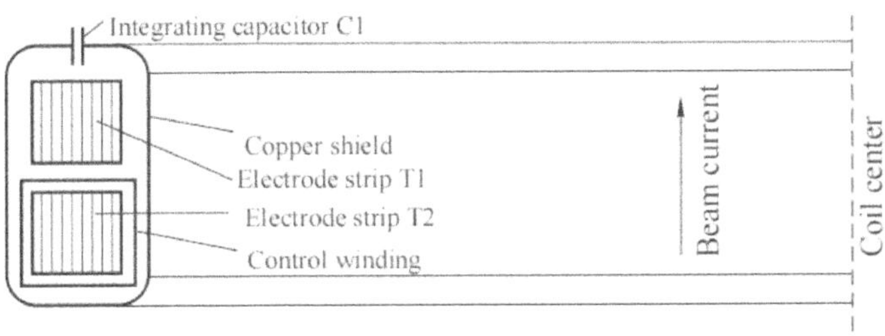

Fig. 9.8 Schematic diagram of ICT power measurement signal.

The beam position monitor (BPM) is used to measure the position and trajectory of the beam in the accelerator. Its main working principle is based on charge collection and signal processing. Typically, the beam position monitor consists of a series of specially structured probes, such as button-type, strip-type, or resonant cavity-type probes. When these detectors are placed on the accelerator track and the beam passes through them, they collect the wakefield signals generated by the beam passing through the electrodes or detectors. The strength of this current or charge signal is proportional to the beam charge and position offset. Subsequently, this current or charge signal is amplified, processed, and analyzed. Common signal processing methods include amplification, filtering, digitization, and data analysis. Through signal processing and analysis, the association between signal amplitude and charge can be eliminated, and the position and trajectory of the beam in the accelerator can be provided. The beam position monitor usually has high precision and fast response characteristics and can monitor and control the beam position in real time. It plays an important role in particle accelerators for optimizing beam trajectories and stabilizing accelerator operation.

Ionization chambers are commonly used in beam diagnostic systems to obtain real-time measurements of beam dose, position, uniformity, and other important information. The common ionization chamber mainly consists of a collector, a high-voltage electrode, and a medium gas. Ionization chambers can be classified into pulse ionization chambers and integrating ionization chambers according to the measurement method, and into planar or coaxial cylindrical ionization chambers according to the shape of the electrodes. The basic structure is shown in Figure 9.9 [23]. The working principle of the ionization chamber is as follows: charged particles entering the sensitive volume (the region enclosed by the electric field lines through the edge of the collector between the two electrodes) ionize the working gas along their tracks, producing ion pairs. Under the influence of the electric field, the electrons and ions in the sensitive volume drift, forming an ionization current. Since the ionization current is proportional to the beam intensity, the beam intensity can be inferred by measuring this current.

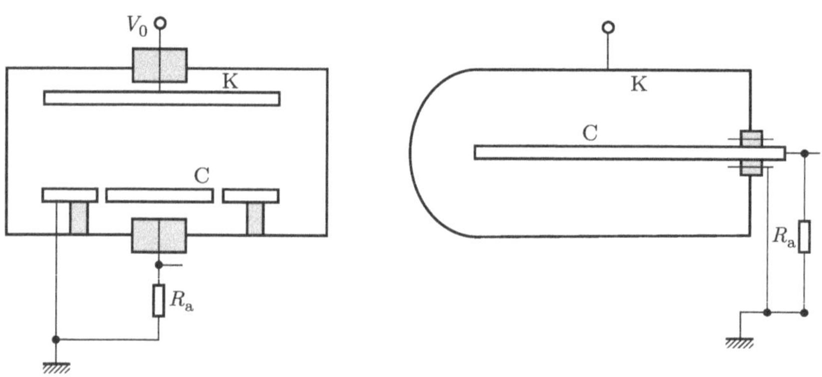

Fig. 9.9 Basic structure of the ionization chamber [23].

9.3 Examples of Beam Devices

Currently, the exploration of laser accelerator facilities worldwide is still in its infancy, with only a few research institutes such as the Extreme Light Infrastructure (ELI) and the Gesellschaft für Schwerionenforschung (GSI) in Germany having built complete laser accelerator beam transport systems. In China, represented by Peking University, a laser accelerator beam transport system with adjustable proton energy (1–15 MeV) and 1% energy spread was constructed in 2018.

The overall layout of GSI's beam transport system is shown in Figure 9.10. The system uses a pulsed solenoid (SOL in the figure) to collect the proton beam and a radiofrequency cavity (CAV) to rotate the longitudinal phase space (Figure 9.11), followed by a binary quadrupole lens (QD1, QD2) for focusing [24].

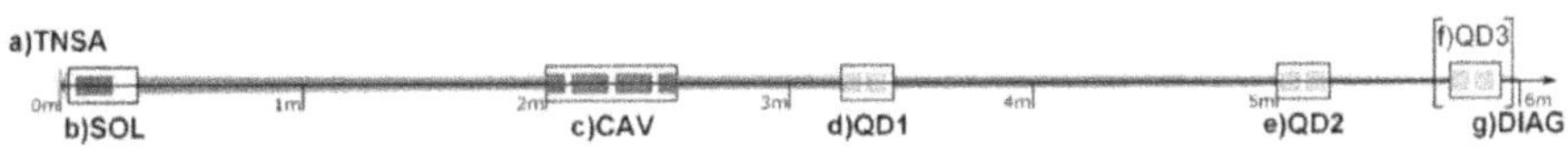

Fig. 9.10 GSI LIGHT beam transport system [24].

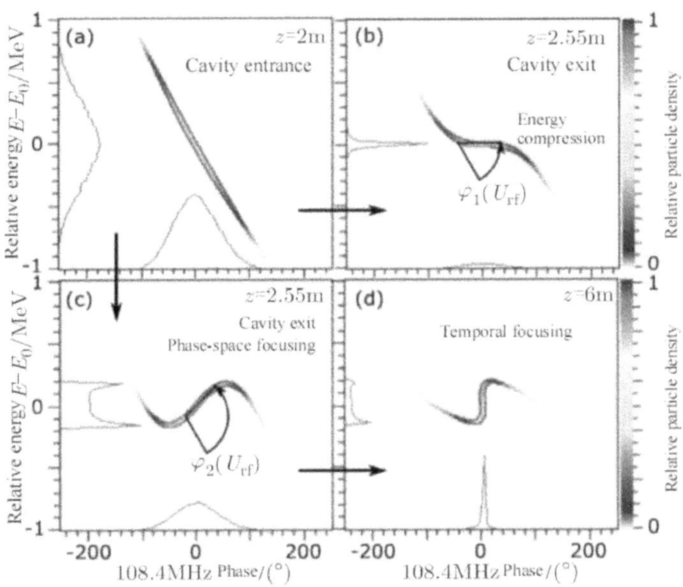

Fig. 9.11 Phase rotation in longitudinal phase space after application of RF cavity [24].

In this GSI system, after focusing and longitudinal phase rotation, a proton bunch with 5×10^8 protons (central energy 7.8 MeV, energy spread $\pm 20\%$) can be compressed to a pulse length of (462 ± 40) ps at a distance of 6 m from the laser target, with a peak current of 170 mA. The focusing elements can effectively increase the collimation of the ion beam, and the radiofrequency cavity can effectively compress the ion beam's pulse length to achieve a high peak current bunch.

The proton transport beam transport system of the European ELI uses quadrupole magnets as the focusing solution. The beam transport system employs five permanent magnet quadrupoles as collecting lenses, a combination of four dipole magnets for energy analysis, and a binary quadrupole lens for focusing. The calculated maximum transport energy is 60 MeV, with an energy spread range of 5%–20% [25].

The United States National Institute for Radiological Sciences, in collaboration with the Helmholtz Institute for Radiation and Nuclear Physics in Germany, proposed a theoretical design for a laser ion accelerator beam transport system for medical applications. In this design, a pulsed solenoid is used to collect the ion beam produced by laser acceleration, focusing is achieved with a solenoid and quadrupole lens, and energy selection is performed with a sector magnet. The design transport energy is approximately 200 MeV, with an energy spread range of 2%–20%, and the output ion beam angle can theoretically rotate 360°, as shown in Figure 9.12 [6].

The CLAPA-I built by Peking University is a laser accelerator capable of producing low-energy spread beams with application value. The facility is based on principles such as Radiation Pressure Acceleration with Phase Stabilization (RPA-PSA) [27, 28] and Target Normal Sheath Acceleration (TNSA) [29], and employs various target materials including metallic targets [30], plastic targets [31], and DLC nano-thin targets [32]. The beam transport system can transport the laser-accelerated ion beam to an irradiation application platform, enabling research in areas such as laser plasma acceleration mechanisms [33], transport of high divergence and strong current ion beams [25], novel target acceleration [34], biological irradiation and medical experiments [35], space radiation environment simulation, and high-energy density physics.

The system uses high-contrast lasers and nm, μm-thick targets to accelerate and produce highly divergent proton beams, which are then collected, analyzed, and reshaped by the beam transport system. As shown in Figure 9.13, the beam transport system consists of three functional sections: (1) The collection section, from the point where the ion beam is produced by laser irradiation to the object point of the deflecting magnet, primarily aims to collect as much of the highly divergent ion beam as possible and focus it at the object point of the deflecting magnet. It mainly includes a ternary quadrupole lens. (2) The energy selection section, from the object

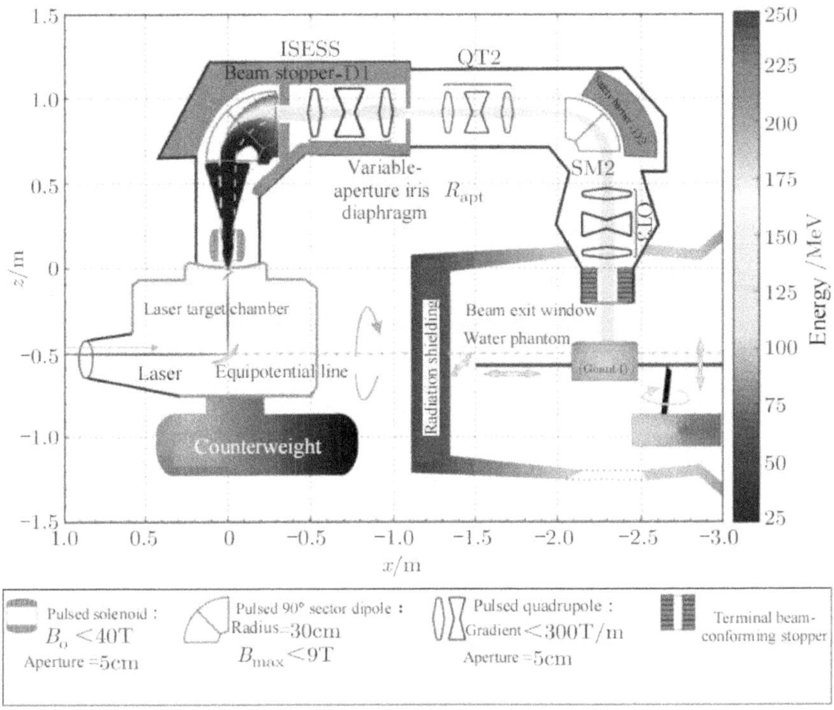

Fig. 9.12 Schematic diagram of transmission device of 360° rotating laser accelerated beam transport system [26].

point to the image point of the deflecting magnet, primarily functions to spatially separate the beam with large energy spread and select the energy through a slit. It mainly includes a 45° sector magnet and two slit systems. Figure 9.14 shows the transverse distribution of the proton beam after passing through the deflecting magnet, with the central energy of the image formed by the deflecting magnet being 5 MeV. It can be seen that in the X direction, beams of different energies spread out, and protons deviating from the central energy cannot be well focused, resulting in a bow-tie shaped beam spot. Therefore, by controlling the size of the slit, the energy spread of the protons can be precisely controlled. (3) The refocusing section, from the image point of the deflecting magnet to the irradiation point at the terminal platform, primarily focuses the beam and fine-tunes its spatial distribution. It mainly includes a binary quadrupole lens. To better monitor and characterize the beam parameters, the entire beam line also includes three beam position detectors: BPD#1 in the collection section,

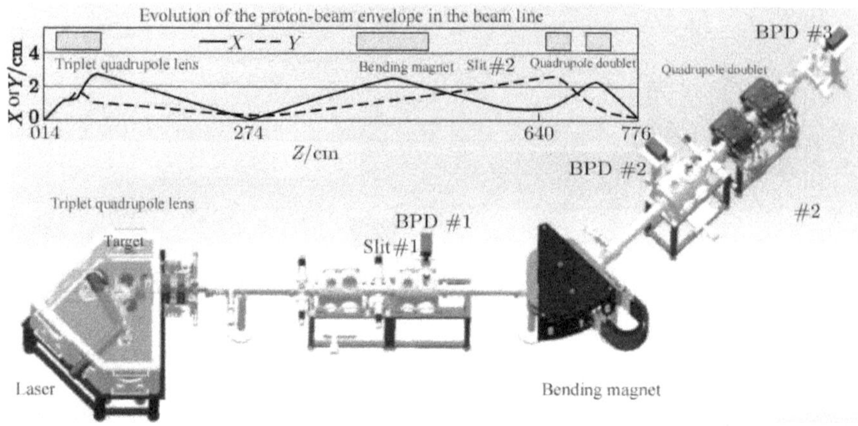

Fig. 9.13 Layout of CLAPA-I beam transport system [39].

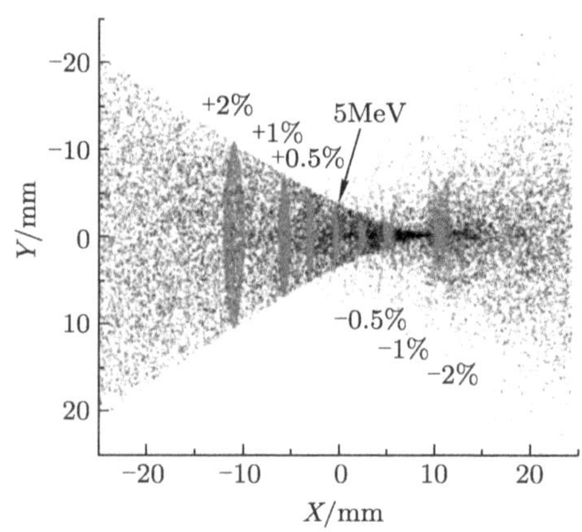

Fig. 9.14 Transverse distribution of proton beam through deflection magnet in CLAPA-I, combined with transverse slit to complete energy selection [39].

BPD#2 in the energy selection section, and MCP-BPD#3 in the irradiation section, located near the focal plane of the ternary quadrupole lens, the image plane of the sector magnet, and the irradiation plane, respectively. Theoretically, the beam line can transport proton beams with energy ranging from 1 to 44 MeV, energy spread ±5%, divergence angle ±50 mrad, and peak current of 8 A.

Figure 9.15 shows the proton charge curve at the irradiation platform (BPD#3) for central energies ranging from 3.5 to 9 MeV. The inset shows

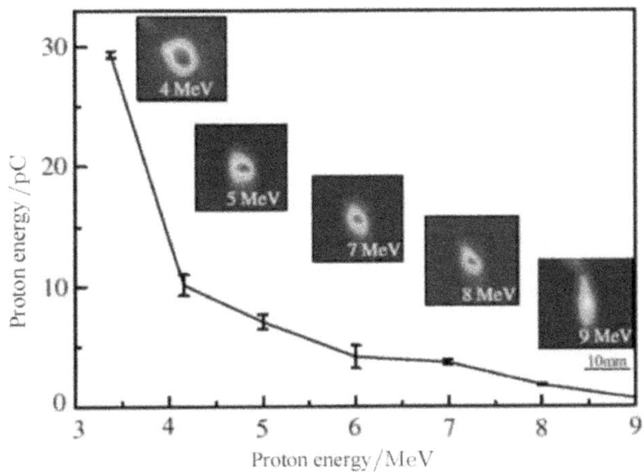

Fig. 9.15 Proton beams with different central energies at the irradiation platform (BPD#3). The images show beam cross sections at different energies. Slit width 14 mm, energy dispersion $\pm 1\%$ [39].

typical beam spot transverse distributions at different central energies, and the beam spot shape and size can be flexibly adjusted by the binary quadrupole lens. Here, the width of the second slit (slit #2) is 14 mm, ensuring an energy spread of $\pm 1\%$.

Figure 9.16 shows the overall layout of the beam delivery system for the CLAPA-II under construction at Peking University. The system is divided into three areas, corresponding to: the common emission degree selection beam collection system, the horizontal terminal beam delivery system (including the beam transport line and treatment head), and the vertical terminal beam delivery system (including the beam transport line and treatment head). Both the horizontal and vertical terminal delivery systems have three functions: energy spread selection, matching transport, and terminal spot control and adjustment [36].

In summary, beam transport systems can achieve precise regulation of laser-accelerated ion beams in terms of energy, charge, and beam spot, ensuring the stability and reliability of the beams. By integrating focusing and energy selection elements reasonably, the inherent large energy spread and large divergence angle of laser-driven beams can be overcome, giving laser accelerators application value. By precisely controlling proton beam parameters, such as energy, energy bandwidth, uniformity, and beam spot size, monoenergetic proton beams can be obtained. Monoenergetic proton beams can be applied in fields such as precise biological dose deposition,

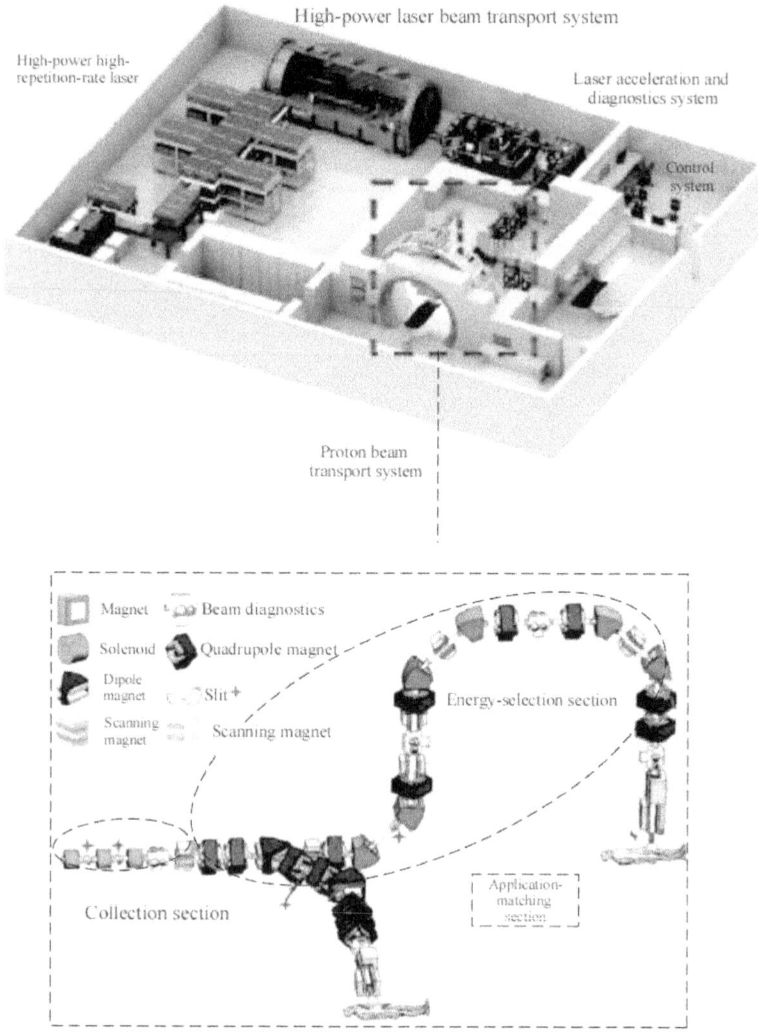

Fig. 9.16 Overall layout of CLAPA-II beam distribution system.

space irradiation environment simulation, measurement of energy stopping in warm dense matter, calibration of detection devices, and proton beam parameter measurement. Designing beam lines for hundred-MeV proton beams and achieving precise regulation of beam parameters and accurate dose delivery will provide feasibility for the next generation of compact radiotherapy devices based on laser accelerators.

Although the technologies used in laser accelerator beam transport systems are mostly mature techniques widely used in traditional accelerators, the characteristics of laser-accelerated beams and the new problems accompanying these characteristics remain new challenges. Currently, research on laser accelerator beam transport systems is still in its infancy and requires continuous exploration.

References

[1] Migliorati M., Bacci A., Benedetti C., *et al.* Intrinsic normalized emittance growth in laser-driven electron accelerators [J]. *Physical Review Special Topics - Accelerators and Beams*, 2013, 16(1): 011302.
[2] Floettmann K. Some basic features of the beam emittance [J]. *Physical Review Special Topics - Accelerators and Beams*, 2003, 6(3): 034202.
[3] Wu M., Zhu J., Li D., *et al.* Collection and focusing of laser accelerated proton beam by an electromagnetic quadrupole triplet lens [J]. *Nuclear Instruments and Methods in Physics Research, Section A: Accelerators, Spectrometers, Detectors and Associated Equipment*, 2020, 955(2020): 163249.
[4] Wu M. J., Li D. Y., Zhu J. G., *et al.* Emittance measurement along transport beam line for laser driven protons [J]. *Physical Review Accelerators and Beams*, 2020, 23(3): 031302.
[5] Chen J. Fundamentals of Accelerator Physics [M]. Beijing: Peking University Press, 2012.
[6] Lee S. Y. Accelerator Physics [M]. Singapore: World Scientific Publishing Company, 2018.
[7] Hofmann I., Meyer-Ter-Vehn J., Yan X., *et al.* Collection and focusing of laser accelerated ion beams for therapy applications [J]. *Physical Review Special Topics Accelerators & Beams*, 2011, 14(3): 031304.
[8] Harres K., Alber I., Tauschwitz, *et al.* Beam collimation and transport of quasi-neutral laser-accelerated protons by a solenoid field [J]. *Journal of Physics Conference Series*, 2010, 244(2): 022036.
[9] Agosteo S., Anania M. P., Caresana M., *et al.* The LILIA (Light Ions laser Induced Acceleration) experiment at LNF [J]. *Nuclear Instruments & Methods in Physics Research*, 2013, 331(331): 15–19.
[10] Burris-Mog T., Harres K., Nürnberg F., *et al.* Laser accelerated protons captured and transported by a pulse power solenoid [J]. *Physical Review Special Topics Accelerators and Beams*, 2011, 14(12): 7–13.
[11] Dewa H., Ahn H., Harano H., *et al.* Experiments of high energy gain laser wakefield acceleration [J]. *Nuclear Instruments & Methods in Physics Research*, 1998, 410(3): 357–363.
[12] Weingartner R., Raith S., Popp A., *et al.* Ultralow emittance electron beams from a laser-wakefield accelerator [J]. *Physical Review Accelerators and Beams*, 2012, 15(11): 111302–111302.
[13] Eichner T., Grüner F., Becker S., *et al.* Miniature magnetic devices for

laser-based, table-top free-electron lasers [J]. *Physrevst Accelbeams*, 2007, 10(8): 330–338.

[14] Osterhoff J., Sokollik T., Nakamura K., *et al.* Transport and non-invasive position detection of electron beams from laser-plasma accelerators [J]. *AIP Conference Proceedings*, 2010, 1299(1): 575–579.

[15] Schollmeier M., Becker S., Geissel M., *et al.* Controlled transport and focusing of laser accelerated protons with miniature magnetic devices [J]. *Physical Review Letters*, 2008, 101(5): 055004.

[16] Nishiuchi M., Daito I., Ikegami M., *et al.* Focusing and spectral enhancement of a repetition-rated, laser-driven, divergent multi-MeV proton beam using permanent quadrupole magnets [J]. *Applied Physics Letters*, 2009, 94(6): 670.

[17] Roth M., Alber I., Bagnoud V., *et al.* Proton acceleration experiments and warm dense matter research using high power lasers [J]. *Plasma Physics and Controlled Fusion*, 2009, 51(12): 124039.

[18] Hofmann I., Jürgen M.-t.-V., Yan X., *et al.* Chromatic energy filter and characterization of laser-accelerated proton beams for particle therapy [J]. *Nuclear Inst & Methods in Physics Research A*, 2012, 681(none): 44–54.

[19] Teng J., Gu Y. Q., Zhu B., *et al.* Beam collimation and energy spectrum compression of laser-accelerated proton beams using solenoid field and RF cavity [J]. *Nuclear Instruments and Methods in Physics Research, Section A: Accelerators, Spectrometers, Detectors and Associated Equipment*, 2013, 729: 399–403.

[20] Hofmann K. M., Schell S., Wilkens J. J. Laser-driven beam lines for delivering intensity modulated radiation therapy with particle beams [J]. *Journal of Biophotonics*, 2012, 5(11–12): 903–911.

[21] Schell S., Wilkens J. J. Modifying proton fluence spectra to generate spread-out Bragg peaks with laser accelerated proton beams [J]. *Physics in Medicine & Biology*, 2009, 54(19): N459–N466.

[22] Nishiuchi M., Sakaki H., Hori T., *et al.* Measured and simulated transport of 1.9 MeV laser-accelerated proton bunches through an integrated test beam line at 1 Hz [J]. *Physical Review Special Topics - Accelerators and Beams*, 2010, 13(7): 071304.

[23] An J., Qing S. Y., Wu H. F. Gaseous Ionization Chamber [M]. Beijing: Atomic Energy Press, 1997.

[24] Busold S., Schumacher D., Brabetz C., *et al.* Towards highest peak intensities for ultra-short MeV-range ion bunches [J]. *Scientific Reports*, 2015, 5: 12459.

[25] Romano F., Schillaci F., Cirrone G. A. P., *et al.* The ELIMED transport and dosimetry beamline for laser-driven ion beams [J]. *Nuclear Instruments and Methods in Physics Research Section A: Accelerators, Spectrometers, Detectors and Associated Equipment*, 2016, 829(1): 153–158.

[26] Masood U., Bussmann M., Cowan T. E., *et al.* A compact solution for ion beam therapy with laser accelerated protons [J]. *Applied Physics B*, 2014, 117(1): 41–52.

[27] Yan X. Q., Lin C., Sheng Z., *et al.* Generating high-current monoenergetic

proton beams by a circularly polarized laser pulse in the phase-stable acceleration regime [J]. *Physical Review Letters*, 2008, 100(13): 135003.
[28] Yan X. Q., Wu H. C., Sheng Z. M., *et al.* Self-organizing GeV, nano-Coulomb, collimated proton beam from laser foil interaction at 7×10^{21} W/cm^2 [J]. *Physical Review Letters*, 2009, 103(13): 135001.
[29] Tudisco S., Altana C., Lanzalone G., *et al.* Investigation on target normal sheath acceleration through measurements of ions energy distribution [J]. *Review of Scientific Instruments*, 2016, 87(2): 02A909.
[30] Torrisi L., Cutroneo M., Ceccio G., *et al.* Near monochromatic 20 MeV proton acceleration using fs laser irradiating Au foils in target normal sheath acceleration regime [J]. *Physics of Plasmas*, 2016, 23(4): 056706–056356.
[31] Torrisi L. Advanced polymer targets for TNSA regime producing 6 MeV protons at 10^{16} W/cm^2 laser intensity [J]. *Physics of Plasmas*, 2017, 24(2): 023111.
[32] Henig A., Kiefer D., Markey K., *et al.* Enhanced laser-driven ion acceleration in the relativistic transparency regime [J]. *Physical Review Letters*, 2009, 103(4): 045002.
[33] Bulanov S. S., Esarey E., Schroeder C. B., *et al.* Radiation pressure acceleration: the factors limiting maximum attainable ion energy [J]. *Physics of Plasmas*, 2016, 23(5): 056703.
[34] Bin J. H., Ma W. J., Wang H. Y., *et al.* Ion acceleration using relativistic pulse shaping in near-critical-density plasmas [J]. *Physical Review Letters*, 2015, 115(6): 064801.
[35] Kraft S. D., Richter C., Zeil K., *et al.* Dose-dependent biological damage of tumor cells by laser-accelerated proton beams [J]. *New Journal of Physics*, 2010, 12(8): 85003–85012.
[36] Wang K. D., Zhu K., Easton M. J., *et al.* Achromatic beamline design for a laser driven proton therapy accelerator [J]. *Physical Review Accelerators and Beams*, 2020, 23(11): 111302.
[37] Migliorati M., Bacci A., Benedetti C., *et al.* Intrinsic normalized emittance growth in laser-driven electron accelerators [J]. *Physical Review Special Topics - Accelerators and Beams*, 2013, 16(1): 011302.
[38] Berz M., Makino K., Wan W. An Introduction to Beam Physics [M]. Boca Raton: Taylor & Francis, 2015.
[39] Zhu J. G., Wu M. J., Liao Q., *et al.* Experimental demonstration of a laser proton accelerator with accurate beam control through image-relaying transport [J]. *Physical Review Accelerators and Beams*, 2019, 22(6): 061302.

Chapter 10

An Introduction to Laser Acceleration Applications

10.1 Applications of Laser-Ion Beams

Laser-ion beams generated from the interaction of ultra-intense lasers and solid targets have several characteristics, such as high instantaneous current density, small size, narrow pulse width, and a wide energy spectrum. These features make them highly promising for a wide range of applications, most of which have been experimentally verified.

10.1.1 *Magnetic Confinement Fusion Diagnostics*

With the steady progress of the International Thermonuclear Experimental Reactor (ITER) project, significant advancements have been made in magnetic confinement fusion research in recent years. This progress has posed new challenges for diagnostics of fusion plasmas, such as the need to develop new methods for diagnosing electrostatic fluctuations in the core plasma and the poloidal magnetic field. The development of novel ion beam probe diagnostic techniques that can be applied to large tokamaks and future fusion research facilities is of great significance, not only for domestic large-scale devices (such as EAST, HL-2A/2M, etc.) but also for international large-scale tokamak devices like ITER.

In magnetic confinement fusion research, high-energy ion beams have long been an important tool for plasma diagnostics. The Heavy Ion Beam Probe (HIBP) utilizes the secondary ionization effect caused by the interaction of heavy ions with plasma to diagnose plasma potential and electron density fluctuations. It is currently the only method capable of measuring the potential and electron density fluctuations in the core plasma. Traditional HIBPs rely on conventional accelerators to produce high-energy heavy ion beams, which are large in size, difficult to construct and maintain, and costly. Moreover, they require large-aperture vacuum chamber windows, severely limiting their application in large tokamak devices.

Using laser-accelerated ion beams for diagnosing the magnetic field and potential of tokamaks could significantly reduce the space, operation, and maintenance costs required for hardware. Additionally, lasers have the flexibility to accelerate ions directly from the inner wall of the vacuum chamber, overcoming the disadvantage of traditional HIBPs that require large-aperture windows. By measuring the toroidal and poloidal displacements of the ion beam exit positions, it is possible to simultaneously diagnose the spatial distribution of the poloidal magnetic field and the radial electric field. Moreover, laser-accelerated ion beams have the advantages of short pulses (on the ps scale), wide energy spread, and multiple charge states. Therefore, when traditional HIBP detection and analysis methods are no longer applicable, developing new diagnostic principles and inversion methods based on laser accelerators is a viable option.

Figure 10.1 illustrates the basic principle of the Laser Ion Beam Track Probe (LITP). Lasers can accelerate proton beams to the MeV level over a distance of μm, allowing the target to be placed near the inner wall of the tokamak device window to generate proton beams. The injection part only requires optical elements, significantly reducing the size requirements for the entry window. More importantly, the wide energy spread of laser-driven ion beams, which is considered a disadvantage, can be converted into an advantage for large-area lateral scanning of the plasma core region, providing the possibility for two-dimensional electromagnetic field diagnostics.

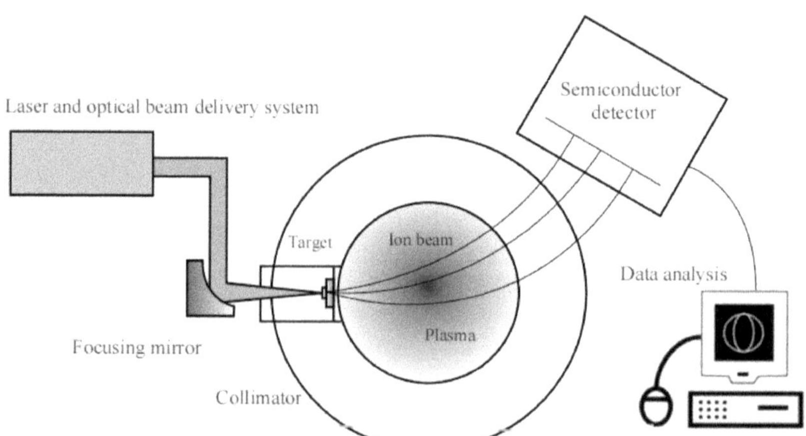

Fig. 10.1 Schematic diagram of the Laser Ion Beam Trajectory Probe (LITP) system.

The short-pulse feature of laser acceleration can be used to measure ion energy through time-of-flight measurements, and combining this with spatiotemporal resolution detectors can yield two-dimensional electromagnetic field diagnostics. The multi-charge state feature can separate the effects of electric and magnetic fields, enabling simultaneous detection of both. Compared to traditional diagnostic methods, the LITP device is smaller in size, easier to maintain, and requires a smaller entry window, making it potentially applicable to large superconducting tokamak systems, such as the ITER device.

In recent years, research on LITP has made significant progress. Researchers from Peking University, including Xiao Chijie and Lin Chen, have derived and improved the diagnostic principles of LITP, controlling the error within 10%. They have also derived the algorithm for two-dimensional diagnostics of the poloidal magnetic field. Moreover, the spatial distribution information of laser heavy ion beams can be used to achieve the same effect as HIBP-obtaining a two-dimensional electron density distribution. By extending the LITP algorithm with nonlinear assumptions, Peking University has verified the LITP diagnostic method on devices with small poloidal magnetic fields, proving its theoretical feasibility for application in spherical tokamaks and compact fusion devices.

LITP represents the first practical application of laser ion acceleration technology in magnetic confinement fusion plasmas. It leverages the unique features of laser-accelerated ion beams and is expected to provide two-dimensional distribution measurements of the poloidal magnetic field, thereby directly obtaining the plasma q-profile and magnetic field configuration. LITP can also perform two-dimensional imaging detection of the radial electric field and electron density, which can be used to study important tokamak physics processes such as plasma turbulence and transport.

10.1.2 *Generation of Warm Dense Matter*

Warm Dense Matter (WDM) is a new form of matter distinct from solid, liquid, gas, and classical plasma. Research on the properties of WDM is of significant scientific value in fields such as inertial confinement fusion (ICF), heavy ion fusion, and astrophysical evolution, as it represents an intermediate state that matter must go through in these processes. As shown in Figure 10.2, the state of WDM lies between solid and classical plasma. Typically, its particle number density ranges from 10^{22} to $10^{25}/\text{cm}^3$, and its temperature ranges from 0.1 to 100 eV.

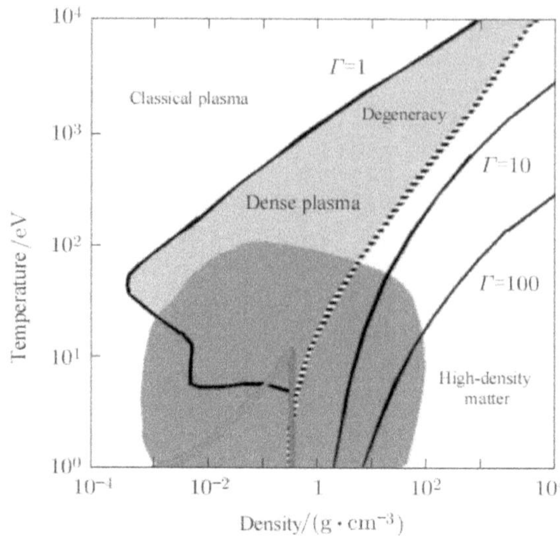

Fig. 10.2 Distribution of warm dense matter on the temperature-density phase diagram.

Currently, several methods are used to generate WDM in the laboratory, including mechanical collision compression (such as flyer plates, gas guns, etc.), chemical energetic material explosion compression, laser pulse irradiation, ion beam heating, and static high-pressure diamond anvil cell (DAC) technology. With the development of laser-plasma acceleration technology, high-energy short-pulse ion beams have gradually been applied to the generation of WDM. Compared to traditional methods, laser ion beams offer the advantages of short heating times and good uniformity of sample states (due to the strong penetration of ion beams and uniform energy deposition within the sample), providing a new means for studying the equation of state of WDM.

In 2003, Patel *et al.* conducted an experiment at the 100 TW laser facility of the Lawrence Livermore National Laboratory in the United States to generate warm dense aluminum using proton beams accelerated by laser pulses [1]. They focused a 100 fs, 10 J short pulse through an off-axis parabolic mirror onto a 10 μm thick aluminum foil. The proton beam generated from the rear surface of the aluminum foil was then used to heat another 10 μm thick aluminum foil to produce warm dense aluminum. In the experiment, they compared the plane target and spherical target configurations by replacing the front aluminum foil that generated the proton beam. The experimental results indicated that, under similar conditions, the spherical configuration indeed focused the proton beam, resulting in warmer dense aluminum (20 eV).

In 2013, Gauthier *et al.* conducted a similar experiment at the LULI laboratory in France to study the ionization balance of high-energy carbon ion beams in warm dense matter [2]. As shown in Figure 10.3, they used one laser pulse (B2) to strike a 10 μm thick gold foil to produce a proton beam, which heated a second 100 nm thick aluminum foil to generate warm dense aluminum. Simultaneously, another laser pulse (B1) was used to strike a 1.5 μm thick polyester film to produce a carbon ion beam, which was then obliquely incident on the warm dense aluminum foil from the other side. The proton beam heating resulted in warm dense aluminum with a temperature of approximately 1 eV. The experimental results showed that the warm dense environment of around 1 eV did not significantly affect the ionization balance of carbon ion beams with energies of 0.04 to 0.05 MeV/u.

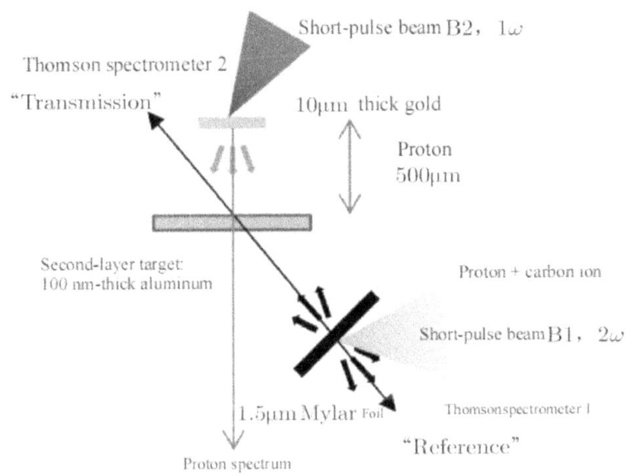

Fig. 10.3 Experimental setup diagram by M. Gauthier *et al.* [2].

10.1.3 *Tumor Therapy*

Proton beams, due to their unique biological energy deposition characteristics, namely the Bragg peak, have significant advantages in tumor therapy. Clinically, physicists can design the physical characteristics of the proton beam (such as energy spectrum structure, beam width, etc.) according to the geometric parameters of the patient's tumor (such as size, shape, position, etc.), so that the Bragg peaks of protons with different energies are deposited in different regions to form an extended Bragg peak (Spread Out Bragg Peak, SOBP). This ultimately deposits most of the proton energy in the target tumor region while minimizing damage to surrounding normal tissue cells. Traditional proton therapy systems based on conventional accelerators face challenges such as high development and maintenance costs and difficulties in widespread adoption. In contrast, laser-driven proton beam systems not only have the advantages of compactness and low cost but also feature short pulse duration and high peak current. Clinically, they can provide ultra-high radiation dose rates and may offer new means for tumor radiotherapy, such as high-dose-rate radiotherapy and immunotherapy.

Radiotherapy with a dose rate greater than 40 Gy/s is referred to as high-dose-rate radiotherapy, also known as FLASH radiotherapy (FLASH-RT). Numerous experimental evidences have shown that high-dose-rate radiotherapy can protect normal tissues to some extent while maintaining the lethality to tumors [3, 4], a phenomenon known as the FLASH effect. This effect is likely related to the consumption of oxygen molecules by free radical chemical reactions-oxygen molecules can fix DNA damage, and a lower concentration of oxygen molecules means less DNA toxicity. However, the effective results of the FLASH effect are currently mostly observed in radiotherapy with electron/photon beams. Due to the strong ionization capability and complex early physicochemical processes of protons/heavy ions, the biological endpoints of FLASH radiotherapy with these particles still need further investigation. Laser-driven proton beams naturally possess the characteristics of short beam duration and ultra-high dose rate, and thus are expected to support the study of this phenomenon.

On the other hand, clinical data indicate that over 90% of cancer patient deaths are caused by cancer cell metastasis [5]. With the current level of cancer diagnosis, less than 10% of clinical patients have not metastasized at the time of diagnosis, and this fortunate group is the target population for the relatively effective traditional three major therapies (surgery, radiotherapy, and chemotherapy). For the other 90% of patients who have experienced local infiltration and distant metastasis, the efficacy of the traditional three major therapies is very limited [6].

In 2018, immunotherapy was awarded the Nobel Prize in Physiology or Medicine for its potential to treat infiltrated and metastasized tumors. With the development of immunotherapy, radiotherapy has gradually expanded from the original 5R theory (5R: Repair, Redistribution,

Reoxygenation, Repopulation, and Radiosensitivity) to the 6R theory, with the addition of: Reactivation of anti-tumor immune response [7].

The phenomenon of radiotherapy activating a systemic, specific immune response was first discovered by Mole in 1953 [8]. He observed that for patients with cancer metastasis, local radiotherapy of the lesion led to systemic tumor growth inhibition. This phenomenon is also known as the abscopal effect. There have been some case reports of clinical cures using the abscopal effect, but its mechanism has been unclear, with speculation that it may be related to various signaling factors released by damaged cancer cells into the bloodstream or other circulatory systems [9]. In the past decade, Peking University has studied the abscopal effect and its partial mechanism between cells and tissues in different biological systems [10–16]. Based on this, a multidisciplinary integrated research method combining physics, chemistry, and clinical medicine can be further used to analyze the key signaling factors released by cancer cells after exposure to pulsed proton beams and their interaction processes and mechanisms with the immune system, with the potential to achieve proton radiotherapy-induced personalized specific "cancer vaccines" [59], thus effectively treating patients with metastasized cancer.

10.1.4 *Ion Acoustic Imaging Technology*

As mentioned above, compared with other cancer radiotherapy methods, ion beam radiotherapy can provide better tumor dose conformity due to the presence of the Bragg peak, protecting healthy organs. However, there is a certain degree of error in predicting the Bragg peak and beam position. Uncertainty in beam range can lead to excessive or insufficient dose deposition in the tumor area, and damage to adjacent healthy tissues [17]. Therefore, it is very important to achieve positioning of the Bragg peak.

The main basis for locating the Bragg peak currently relies on three different types of secondary radiation produced by the interaction of ion beams with biological tissue. The first is positron annihilation γ-rays. When ion beams are incident on tissue, inelastic nuclear interactions may produce isotopes that can emit positrons, and each positron annihilation produces two γ photons. Measuring the γ-ray signal from positron annihilation can achieve positioning of the Bragg peak. Since existing PET scanners [18] can be utilized, this method has been widely studied in clinical settings. The second method is instantaneous γ-rays. Nuclear scattering events cause atomic nuclei to enter an excited state, which then decay to produce instantaneous γ-rays. The advantage of this method is that it is not affected by physiological processes (such as blood flow) and has a larger reaction cross-section.

The third type is ion acoustic imaging technology. The discovery of the ion acoustic effect originates from the theory of thermoacoustic effects, which is now widely applied in photoacoustic imaging technology. The generation of the ion acoustic effect requires irradiating biological tissue with a short-pulse (μs level) ion beam. After absorbing the irradiation energy, the biological tissue locally heats up rapidly and thermally expands, thereby generating ultrasonic waves that propagate outward. The formation and detection of ultrasonic signals during treatment are shown in Figure 10.4. The first two methods utilize γ-photon signals generated by nuclear interactions, which only absorb a small portion of the dose. In contrast, electromagnetic interactions are the main mechanism by which proton beams transfer energy to tissue. Ion acoustic imaging obtains signals from electromagnetic interactions, which are more closely related to the actual dose deposition in the patient's body [19]. It is worth noting that ultrasonic signals attenuate strongly in tissue, so the different effects of various tissues need to be considered [20]. Additionally, ultrasonic signals strongly reflect at the skin boundary; ultrasonic signals cannot be detected in air, and the patient's skin is not always suitable for placing ultrasonic probes. These factors also limit the application of ion acoustic detection.

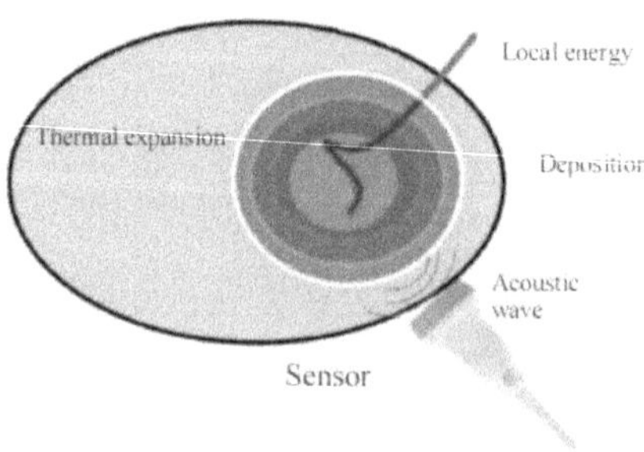

Fig. 10.4 Schematic diagram of the formation and detection of ultrasonic signals during the transmission of monoenergetic ion beams with the Bragg peak characteristic [21].

In 1979, Sulak *et al.* from Harvard University in the United States first irradiated a water phantom with a proton beam exceeding 150 MeV [22], experimentally demonstrating that the instantaneous thermal expansion caused by the proton beam can produce a sufficiently strong detectable ultrasonic signal. In 1995, researchers from the University of Tsukuba in Japan first proposed applying the ion acoustic effect to proton therapy [23]. They detected ultrasonic signals with a hydrophone and used this as the basis to determine the position of the Bragg peak, but the spatial resolution of the ultrasonic signal was very low, only 3 mm.

With the advancement of ultrasonic probe technology and the increasing complexity of radiotherapy transport systems, ion acoustic imaging technology regained momentum after 2013. Teams from the University of Munich in Germany led by Parodi and Assmann [24–27], the University of Pennsylvania in the United States led by Jones [20, 28, 29], and the University of Wisconsin–Madison led by Patch [30, 31] have further explored this field. In 2015, the Munich University team used a 20 MeV proton beam generated by a tandem accelerator [24] to measure the ultrasonic signal at the entrance of the proton beam in water and the ultrasonic echo signal at the Bragg peak, successfully achieving sub-mm precision positioning of the Bragg peak.

During the ultrasonic generation process, the amplitude of the sound wave can only reach its maximum when the ion dose deposition time is less than the time it takes for the sound wave to travel through the Bragg peak. For example, for the lowest energy 70 MeV medical proton beam, the pulse width needs to be less than $5\mu s$ [26]. At this time scale, the energy deposition process can be considered adiabatic, and thermal diffusion can be neglected. Medical cyclotrons typically provide quasi-continuous proton beams, which generally require the addition of choppers [32] or signal modulation techniques [33] to narrow the pulse in the time domain. In contrast, laser ion beams have an inherently short pulse width, with a pulse width in the ns range after emission, allowing ion acoustic imaging without time-domain modulation.

In 2019, the Munich University team first applied ion acoustic measurements to the field of laser acceleration [27], recording the ultrasonic signal generated by the energy deposition of the proton beam in water (as shown in Figure 10.5), and successfully inverted the energy spectrum structure of the ion beam using a simulated annealing algorithm. Ion acoustic detectors can completely monitor a single focused proton beam and provide rapid readouts, while also featuring high repeatability, robustness, and electromagnetic pulse resistance, potentially meeting the experimental and application research needs in the field of laser acceleration in the future.

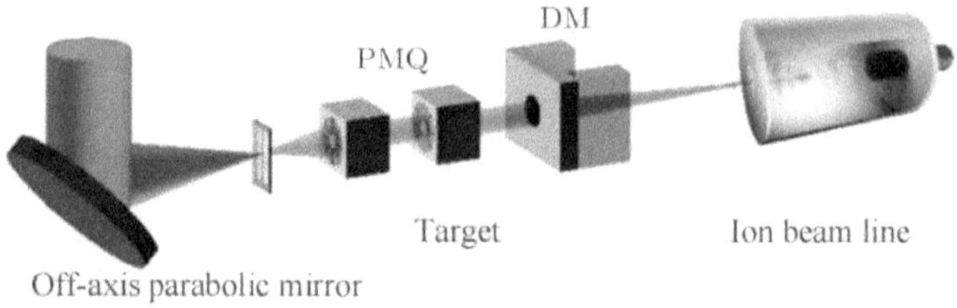

Fig. 10.5 Shows the experimental layout of the ion acoustic detector, where the laser focuses on the target surface to produce a proton beam, which is then focused and energy-selected before entering the detector [27].

10.1.5 *Proton Beam Transmission Bioimaging*

Proton beam transmission imaging, also known as proton radiography, is based on the principle of measuring the attenuation or scattering of the proton beam incident on the object under test to determine the physical properties and geometric structure of the object. It was first proposed by West [56] in 1972. When a high-energy parallel proton beam strikes a material, scattering occurs. The scattered protons have a scattering angle after passing through the material, and there is a difference in intensity compared to the protons that are not irradiated by the material and directly strike the detector, forming a contrast for imaging. Compared with X-ray radiography, proton radiography has the following advantages: protons are sensitive to density and material, and can distinguish between two materials with small density differences-experimental density discrimination of 1% has been achieved; it has strong fine structure resolution capabilities. In addition, unlike X-rays, proton scattering can be controlled. Scattered protons can be focused to form visually background-free, high-contrast images. In contrast, large-angle scattering of the object under test on X-rays cannot be controlled, which reduces the accuracy and sensitivity of radiography.

Figure 1 in Ref. [56] shows the comparison of proton radiography (left) and X-ray radiography (right) results for a mouse sample. The left image is the direct scattering imaging result of a 160 MeV proton beam on a mouse sample, and the right image is the X-ray imaging result. From the comparison, it can be seen that the proton radiography image has very clear edge details and can reveal some details of the object that are not visible with X-ray radiography, which gives proton radiography great application potential in the fields of medical and biological imaging.

For proton radiography applications, compared with the monochromatic parallel proton beam obtained from traditional accelerators, laser-accelerated proton beams have some unique advantages due to their small source size, large divergence angle, and wide energy spectrum. Using a point-like divergent proton source can achieve μm-level spatial resolution. By adjusting the distance between the object to be imaged, the source, and the imaging screen, the magnification of the image can also be controlled. The continuous energy spectrum allows a single shot to irradiate samples with protons of different energies. Placing multiple layers of radiation-sensitive film (RCF) behind the sample can achieve layer-by-layer imaging of protons in multiple energy segments. By detecting the two-dimensional distribution of protons with multiple energies and then inverting the calculation, more abundant and accurate density and structural information of biological tissue can be obtained.

Internationally and domestically, some research groups have carried out preliminary studies on biological transmission imaging using laser proton beams. For example, a research team from the Shanghai Institute of Optics and Fine Mechanics conducted wide-spectrum transmission imaging of a dragonfly using a double-layer RCF. Figure 10.6 shows the experimental setup and results. The image on the first layer of RCF comes from protons with energies of 0.5 to 1 MeV, and the image on the second layer of RCF comes from protons with energies greater than 2.8 MeV [57].

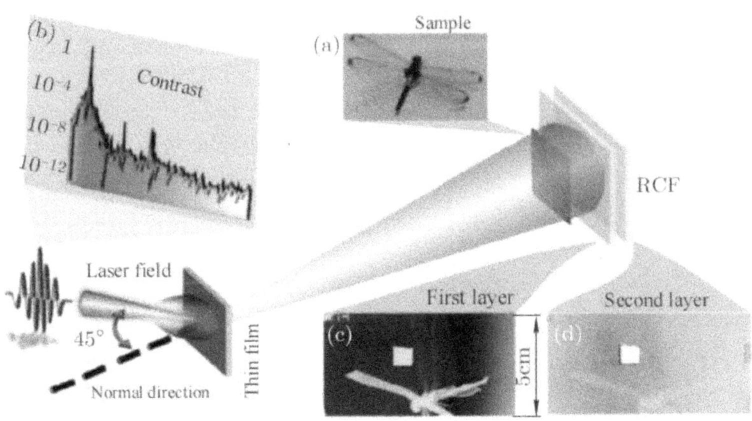

Fig. 10.6 Imaging of a dragonfly using a laser-accelerated broad-spectrum proton beam [57].

Figure 10.7 shows a partial enlarged view of the dragonfly structure. It can be seen that proton imaging is very sensitive to subtle density changes within the dragonfly and has high contrast, with a lateral resolution reaching the μm level.

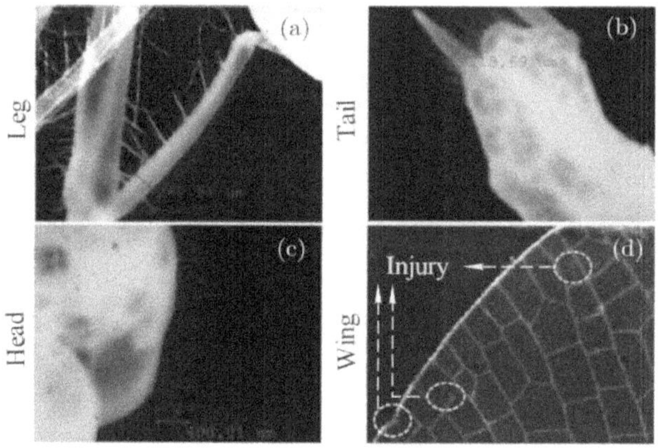

Fig. 10.7 Close-up images of a dragonfly imaged with a broad-spectrum proton beam [57]. (a) Leg, (b) tail, (c) head, (d) wing.

10.1.6 *High-Density Plasma Electromagnetic Field Diagnostics*

Laser-accelerated proton beams have the characteristic of a wide energy spectrum, and different energy proton beams have different flight times from the point source to the detection area. Combined with the use of multi-layer radiation-sensitive film (RCF), the energy distribution of the proton beam can be converted into a time distribution in the detection area, thereby capturing the physical processes at different moments. Proton beams are an important tool for temporally and spatially resolved diagnostics of electromagnetic fields in laser-plasma interactions.

In 2002, Borghesi [34] first used laser-accelerated proton beams to diagnose the electromagnetic fields in laser-plasma interactions. The experimental setup is shown in Figure 10.8. One laser pulse (CPA_1) is used to generate the proton beam, and another laser pulse (CPA_2) interacts with matter to produce a high-intensity transient electromagnetic field. The detector consists of a stack of RCFs. When the proton beam deposits energy in the RCFs, the color of the RCFs changes, thereby obtaining the

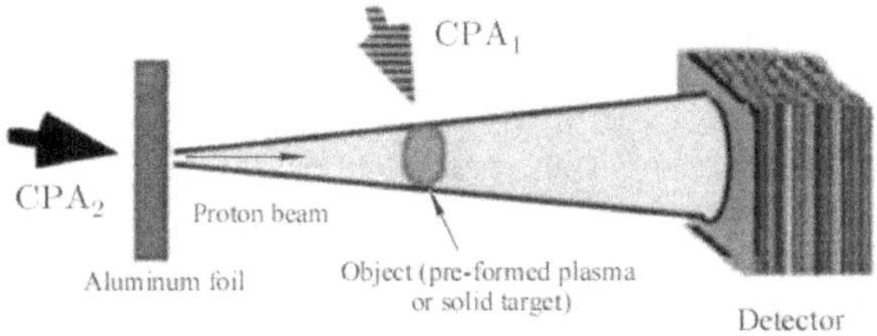

Fig. 10.8 Shows the proton radiography experimental setup: one laser pulse (CPA_1) is used to generate the proton beam, and another laser pulse (CPA_2) interacts with matter to produce a high-intensity transient electromagnetic field [34].

average dose of the deposited proton beam. If multiple RCFs are stacked together, by measuring the dose of protons deposited in each film, the dose and energy distribution of the proton beam can be obtained. Figure 3 in Ref. [34] shows the laser pulse (CPA_1) interaction with a 50 μm thick Ta wire captured by 6 to 7 MeV protons at different delay times ((a) −15 ps, (b) −5 ps, (c) 5 ps, (d) 15 ps, (e) 25 ps). From the proton radiography images, the influence of the electromagnetic field on the proton beam as it passes through the plasma region can be inferred, thereby obtaining the time evolution information of the transient electromagnetic field distribution.

In 2016, Kar *et al.* [35] explored the dynamics of hot electron escape in the TNSA acceleration mechanism. They utilized the wide energy spectrum of protons and the multi-layer radiography capability of RCF to perform a series of dynamic imaging of the hot electron escape process using a single laser beam. One of the experimental setups and results is shown in Figure 10.9. During the interaction of the target with the laser, a large number of escaping electrons were generated, causing the target to become positively charged. To restore the electrical neutrality of the target, the positive charge within the target flowed along the metal wire towards the ground. Using a laser-accelerated proton beam with a large energy spread, the induced electric field of the current at different moments was captured at (b) 101 ps, (c) 116 ps, (d) 153 ps. The electric field was inverted using a program to obtain the distribution of the electromagnetic field in the metal wire — that is, the charge line density distribution of the metal wire under the action of the field. After integration calculation, the total charge in the pulse was found to be approximately 60 nC.

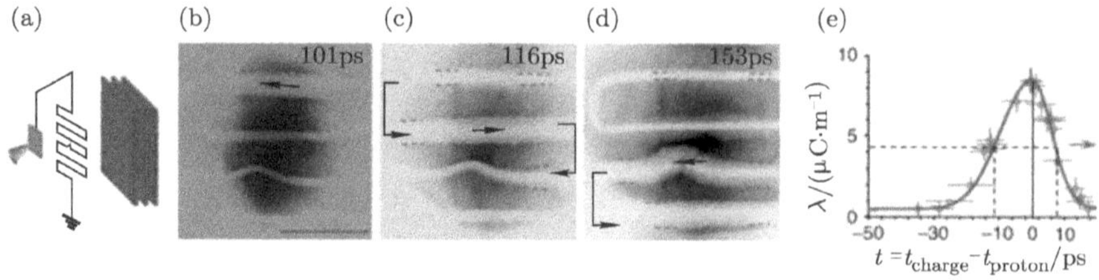

Fig. 10.9 During the interaction between the target and the laser, a large number of electrons escape, causing the target to become positively charged. To restore electrical neutrality, the positive charges within the target flow along the metal wire to the ground. Using a laser-accelerated proton beam with large energy spread, proton beams of different energies captured the current-induced electromagnetic field at different moments: (b) 101 ps, (c) 116 ps, (d) 153 ps. (e) Shows the distribution of the electromagnetic field in the metal wire obtained through electric field inversion using a program [35].

10.1.7 *Applications in Material Science*

In recent years, laser-accelerated proton beams have found widespread applications in material science [36–39]. Due to their high instantaneous current density and short pulse duration, laser-accelerated proton beams have a significant thermal effect, capable of heating samples to very high temperatures. As the irradiation distance decreases, the heating temperature gradually increases [36, 37]. Simulation results indicate that the thermal effect primarily originates from the laser-accelerated proton beam, with lesser contributions from electrons, photons, and heavy ions [37]. Compared to continuous beam irradiation from traditional accelerators, laser-accelerated pulsed proton beams possess distinct characteristics.

Laser-accelerated proton beams can be used for high-precision synthesis of nm- and μm-scale crystalline particles [38, 39]. In material science, the synthesis and growth of nm- and μm-scale crystalline particles is a frontier area facing significant challenges, with applications in multiple fields. The main factors determining the growth of these crystals are the growth temperature and pressure conditions, which need to be achieved within a very short time range (ps-ns). Traditional techniques currently lack sufficient precision in controlling these conditions, whereas laser-driven short-pulse, high-current proton beams can meet the temperature and pressure requirements for crystal structure growth.

For example, as shown in Figure 10.10, an experiment at the TITAN laser facility at Lawrence Livermore National Laboratory in the United States used a laser pulse width of 700 ps and a laser intensity of 10^{20} W/cm^2. The accelerated protons struck a gold target located 2.5 cm behind the

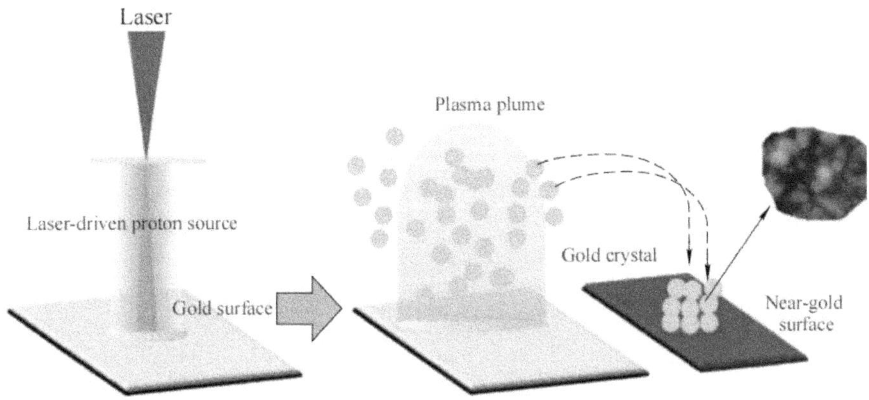

Fig. 10.10 High-precision synthesis of nm and μm scale crystal particles using laser-accelerated proton beams.

target, with a silver target receiving the gold microcrystalline particles. The laser-accelerated proton beam heated the gold target to a temperature between the boiling point and critical point within a few hundred ps. The plasma plume (i.e., atoms, ions, nanoclusters, etc.) generated by the ablation of the material expanded into vacuum and cooled down. The plasma plume nucleated and formed microcrystalline structures on the surrounding cold surface. Compared with traditional methods, the advantage of laser-accelerated proton beam synthesis of microcrystalline structures is higher speed and precision; however, the disadvantage is that the synthesized micro- and nanocrystals need to be separated from the surface, and additional steps are required before the nanocrystals can be used, increasing complexity.

Laser proton beams can also be used for testing the radiation resistance of materials [37]. The high instantaneous current density of laser-accelerated proton beams provides ideal conditions for testing materials that operate under extreme conditions, such as high-energy density physics, astrophysics, space radiation environments, and plasma extreme environments in magnetic confinement fusion and inertial confinement fusion. Compared with traditional methods of testing the radiation resistance of materials using accelerators, laser-accelerated proton beams have short pulse widths and high instantaneous current densities. Single-shot laser-accelerated proton beams can produce transient thermal shocks for material testing under extreme conditions, significantly reducing experimental time and cost. The materials of interest in this field are typically high-melting-point materials required for fusion devices (e.g., tungsten), as there is a strong demand to improve the radiation resistance of these materials.

Figure 10.11 shows the surface morphology of metal tungsten before and after irradiation with laser-accelerated proton beams. The initially smooth sample surface shows cracks and pores, indicating strong surface erosion caused by laser-accelerated proton beam irradiation. This result has very similar characteristics to those obtained using traditional methods, proving that laser-accelerated proton beams can be used for testing the radiation resistance of materials under extreme conditions, just like traditional accelerators.

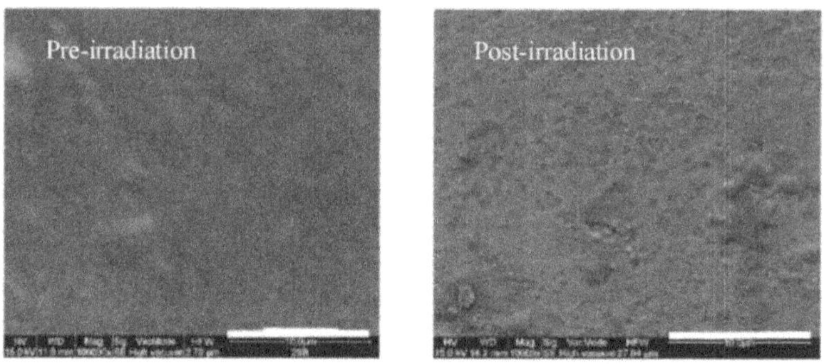

Fig. 10.11 Electron microscope images of the surface morphology of metallic tungsten before and after irradiation, using laser-accelerated proton beams for testing the radiation resistance of materials under extreme conditions.

10.1.8 *Inertial Fusion Proton Fast Ignition*

Controlled nuclear fusion is one of the most effective ways to solve the energy crisis in the future. Obtaining more sustainable and safer energy through nuclear fusion is a long-term strategic goal for humanity. Currently, there are two main technological approaches to nuclear fusion: magnetic confinement and inertial confinement, both of which are still in the exploratory stage. For inertial confinement fusion, when the energy released by fusion is higher than the energy consumption due to processes such as bremsstrahlung and heat diffusion, the fuel can burn self-sustainably. The process of achieving this state is called ignition, which is a key step in the inertial confinement nuclear fusion process.

The traditional ignition method for inertial confinement fusion involves using multiple laser beams to directly irradiate or indirectly irradiate a deuterium-tritium (DT) target sphere with uniform X-rays generated by irradiating the inner wall of a high-Z material cavity, producing a rapid, highly spherical implosion–compression–hotspot to achieve ignition, requiring laser output energy in the MJ range. This scheme has extremely high requirements for the sphericity and uniformity of laser irradiation — if the laser focus spot is uneven or the target surface is not smooth enough, Rayleigh-Taylor (RT) instability may occur. This non-uniformity will be exponentially amplified during the compression process, leading to ignition failure. Therefore, to overcome the problems in traditional ignition schemes, the concept of "fast ignition" was proposed. The fast ignition scheme can reduce the demand for driving laser energy, and the compression and ignition processes are separated, which can also greatly reduce the requirements for symmetry during the compression stage. Ultra-hot electrons or high-energy proton beams generated by the interaction of ultra-intense lasers with plasma can serve as the particle sources for driving fast ignition.

In 1994, Tabak *et al.* from the Lawrence Livermore National Laboratory in the United States [40] proposed a scheme for fast ignition using ultra-hot electrons generated by ultra-intense lasers. First, a hollow target sphere is highly symmetrically compressed with a medium-strength long-pulse laser (main pulse duration of several ns, wavelength of 0.35 μm, $I = 10^{14} \sim 10^{15}$ W/cm^2). The target sphere is filled with deuterium-tritium gas, and after compression, the center is a high-density region ($\rho = 300$ g/cm^3), while the outside is a high-temperature, low-density plasma corona region several mm to cm long. The implosion speed in this process is relatively low. Then, a laser pulse with a pulse width of 100 ps is incident from the corona to a location about 100 times the critical density, further compressing the target sphere towards the center and relying on its hole-boring effect to displace material in the corona and form a plasma channel. Finally, an ultra-intense, ultra-short ignition laser ($1 \sim 10$ ps, $\lambda = 1.05$ μm, $I \approx 10^{19}$ W/cm^2) passes through the channel to reach the high-density region, generating a large number of ultra-hot electrons with kinetic energy of about 1 MeV. The energy of the ultra-hot electrons is deposited in the edge of the ignition hotspot in the core region, and the local temperature of the fuel near the target center rapidly rises to the ignition temperature, thereby achieving "fast ignition."

Kodama *et al.* [41] also proposed an experimental configuration for fast ignition using a cone target to guide the injection of the ignition laser. Compared with the configuration proposed by Tabak *et al.*, which uses the hole-boring effect of the laser to form a plasma channel, it allows laser energy to be directly focused near the high-density region. At the same time, the conical structure can also focus the laser and ultra-hot electrons. In 2010, Osaka University in Japan [42] used a gold cone to guide two laser pulses with a pulse width of 1.5 ps to ignite a compressed target, increasing the neutron yield of the thermonuclear reaction from 10^6 without the ps ignition laser to 3.5×10^7. In 2011, the gold cone-guided ignition experiment at the University of Rochester in the United States [43] also achieved a neutron yield four times higher than that without the ignition pulse.

In the process of interaction between ultra-intense lasers and plasma, in addition to generating ultra-hot electrons, high-energy proton beams and heavy ion beams with energy levels of tens of MeV can also be produced. Based on the proton beam generated by the interaction of a laser with a solid thin-film target, which has the characteristics of high energy, good directionality, and high conversion efficiency, and the fact that proton beams can deposit energy mainly at the end of their range, i.e., the existence of a Bragg peak, Roth *et al.* from GSI in Germany proposed the idea of using proton fast ignition for nuclear fusion research in 2000 [58]. The specific layout is shown in Figure 10.12. Outside the indirect-drive hohlraum, multiple PW laser beams are focused on a solid target. The back surface of the target is spherical, which is conducive to focusing the emitted proton beam. The accelerated proton beam then passes through a protective thin metal window and deposits energy in the ignition hotspot, heating it to an ignition temperature exceeding 10 keV. Laser proton beams can be accelerated

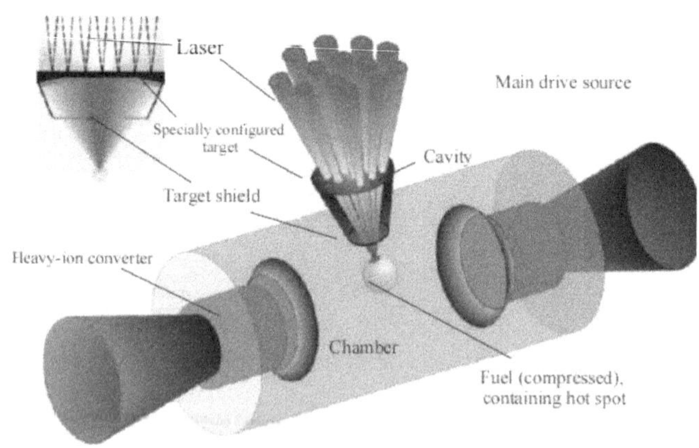

Fig. 10.12 Schematic diagram of the fast ignition device [58].

to tens of MeV on a hundred μm scale, and proton beam focusing can be achieved through target configuration design. Therefore, laser proton beams can be generated near the target sphere and transmitted to it, while proton beams generated by traditional accelerators would encounter many problems in transmission and focusing.

10.1.9 *Applications in Hydrogen-Boron Fusion*

As mentioned in the previous section, laser-driven inertial confinement fusion is an important technology that holds promise for solving energy problems. Currently, laser fusion research represented by the National Ignition Facility (NIF) in the United States mainly adopts an indirect-drive scheme. France is also conducting the Laser Mégajoule (LMJ) research program, and Russia has launched similar projects. However, this indirect-drive fusion method actually has low driving efficiency. Even the highest fusion output energy achieved so far (~26 kJ) is only ~1% of the total laser energy (~1.9 MJ), making it difficult to become a feasible technological approach for generating fusion energy. Laser-constrained fusion aimed at achieving sustainable fusion energy still needs to explore new technological routes [44], which requires solving new physical problems centered on improving efficiency. Summarizing the history of laser fusion research over the past forty years, since the formal proposal of the complete principle of laser inertial confinement fusion, the mainstream scheme has been to use high-power lasers with ns pulse width (laser power density $\sim 10^{15}$ W/cm^2) to symmetrically compress and heat fusion fuel, thereby achieving thermonuclear fusion combustion in a thermal equilibrium state. In the past ten years, the emergence and rapid development of ultra-intense, ultra-short laser pulses (ps or even fs level) have opened up a completely new frontier in the field of relativistic ultra-intense laser plasma physics. Ultra-fast, ultra-intense lasers interacting with non-spherical targets can generate very high plasma temperatures and densities, and the ultra-short pulse makes the fusion reaction far from equilibrium, potentially bringing revolutionary breakthroughs to fusion driving methods.

On the other hand, the current status and development trends of ultra-intense laser and fusion physics research at home and abroad have shown that the hydrogen-boron reaction ^{11}B(p, α)2α is a promising new pathway for fusion reactions. Compared with the traditional fuel DT and DD-induced nuclear fusion reactions, the hydrogen-boron reaction has obvious advantages: the products are high-energy α particles, with virtually no neutron radiation, making it a neutron-free clean fuel; the fusion raw materials are stable and abundant; and the thermonuclear reaction rate in thermal equilibrium is also high, as shown in Figure 1 in Ref. [45], with a value only slightly lower than that of the DT reaction. However, the hydrogen-boron reaction has a higher Coulomb barrier, and thus the energy required for the reaction is also higher. Traditional accelerator experimental data show that the p+^{11}B fusion reaction has two relatively obvious resonance peaks at 140 keV and 660 keV, with cross-sections of about 0.1 barn and 1.2 barn, respectively (1 barn = 10^{-28} m^2), as shown in Figure 1 in Ref. [45]. The cross-section drops sharply below 100 keV at low energy, with an exponential relationship with energy. Therefore, to conduct hydrogen-boron fusion reactions using lasers, the required plasma temperature exceeds 100 keV, which is an order of magnitude higher than that of DD and DT, posing higher requirements for the driving laser power. If the traditional spherical compression inertial confinement fusion mode is still used to drive hydrogen-boron fusion, the bremsstrahlung loss in high-temperature thermal equilibrium will be very large, making it difficult to achieve energy self-sufficiency and gain. Therefore, current research on neutron-free hydrogen-boron fusion is still in its infancy.

Driving hydrogen-boron fusion far from thermal equilibrium can effectively reduce radiation losses and potentially achieve energy self-sufficiency. Several laboratories around the world are conducting related research using strong lasers, all of which basically adopt the configuration of using ultra-intense laser-accelerated proton beams to bombard boron-containing planar targets to drive hydrogen-boron fusion. For example, the LULI laboratory in France conducted a two-beam laser scheme experiment using the Pico2000 facility. One laser beam with an energy of 20 J uses the target normal sheath acceleration (TNSA) ion acceleration mechanism to accelerate and produce a proton beam in the MeV range, which bombards the plasma target generated by the ablation of a boron-containing planar target with another laser beam with an energy of 400 J, achieving a reaction yield of 10^7/sr [46]. In addition, at the PALS laser facility in Prague, Czech Republic, researchers adjusted the pulse shape of a single 600 J long-pulse nanosecond laser to continuously and stably compress and heat a sandwich-structured hydrogen-boron-silicon planar target, producing a hydrogen-boron nuclear reaction. In 2014, they achieved the strongest alpha particle source in the world at that time, reaching 10^9/sr [47]. Recently, they have further achieved a new record yield of 10^{11} alpha particles per laser shot [48]. Even so, the highest fusion energy achieved in these two experiments is less than 0.1 J, which is less than 0.02% of the driving laser energy (600 J). To achieve a breakthrough in laser driving efficiency, further optimization research is needed on the selection of ultra-intense lasers, the physical mechanisms of laser-target interactions, laser driving methods, and target design.

Researching ultra-intense laser-driven hydrogen-boron reactions and exploring and realizing fusion reactions far from thermal equilibrium involves a series of key scientific issues — such as finding ways to achieve fusion reactions assisted by magnetic fields, designing new nanostructured targets, and improving laser conversion efficiency, and even trying various combinations of multiple laser beams (nanosecond, picosecond, femtosecond). These are all of great significance for exploring the ignition conditions of a $^{11}\mathrm{B}(\mathrm{p},\alpha)2\alpha$ reactor driven by lasers. At present, in China, the Shanghai Institute of Optics and Fine Mechanics, Fudan University, and Shanghai Normal University have carried out a large amount of theoretical and experimental research under the support of the CAS Pioneer Project. Fudan University, the China Academy of Atomic Energy, and the Shanghai Advanced Research Institute have also recently conducted preliminary hydrogen-boron/deuterium-deuterium reaction experiments on the CLAPA laser accelerator at Peking University, further enriching the research content of neutron-free nuclear fusion.

10.2 Secondary Radiation in Ion Acceleration Processes

During the process of ultra-intense laser-driven plasma acceleration, various high-brightness secondary radiations in different wavelength bands are often generated. Currently, experiments have been able to produce a variety of secondary radiation sources, such as terahertz radiation [49], mid-infrared pulses [50], extreme ultraviolet radiation [51], K_α radiation [52], X-ray radiation [53], and γ-ray radiation [54], etc. Table 10.1 shows the common parameters of some secondary radiation sources based on ultra-intense laser-plasma interactions. It can be seen that, thanks to the ultra-short and ultra-intense characteristics of the laser, these secondary radiations in different wavelength bands often achieve brightness levels that are difficult for traditional methods to reach. The following sections will briefly introduce several secondary radiations in ion acceleration processes based on solid targets, including terahertz radiation, extreme ultraviolet radiation, and X-ray radiation.

Table 10.1 Common parameters of some secondary radiation sources based on ultra-intense laser-plasma interactions.

Radiation Source	Photon Energy	Photon Number	Brightness*	Divergence Angle	Energy Spectrum Characteristics	Coherence
K_α radiation	~10 keV	$10^6 \sim 10^9$	10^{18}	4π rad	Line spectrum	Incoherent
Betatron radiation	1~30 keV	$10^6 \sim 10^8$	10^{22}	Tens of mrad	Broadband	Spatially coherent
Inverse compton scattering	10 keV~MeV	$10^5 \sim 10^7$	10^{21}	Tens of mrad	Quasi-monoenergetic spectrum	Incoherent
Surface high harmonics	Tens of eV~keV	$10^8 \sim 10^{12}$	10^{28}	Tens of mrad	Broadband	Temporally and spatially coherent
Terahertz radiation	$10^{-4} \sim 0.1$ eV	$10^{14} \sim 10^{17}$	10^{25}	~1 rad	Broadband	Temporally and spatially coherent

*Brightness unit: Photons/s/mm^2/mrad^2/0.1%BW.

10.2.1 *Terahertz Radiation*

Terahertz (THz) radiation generally refers to electromagnetic radiation in the frequency range of 0.1 THz to tens of THz, located in the spectrum between microwave and far-infrared bands. Unlike microwave and far-infrared bands, terahertz radiation has many unique properties, such as different spectral characteristics and absorption/transmission properties for different types of materials. These properties make terahertz radiation widely applicable in fields such as imaging, material detection, communication, and material science. Unlike its adjacent bands, the terahertz range has long been lacking in high-power radiation sources. Currently, there are two mainstream approaches to generating terahertz radiation: one based on ultrafast femtosecond lasers and the other based on ultra-short electron beams.

The representative method of terahertz radiation sources based on ultrafast lasers is the photoconductive antenna method and optical rectification method. In the photoconductive antenna method, semiconductors generate photoexcited carriers under the action of ultrafast lasers, which then form transient currents under the influence of a bias voltage, thereby radiating electromagnetic waves outward. The optical rectification method utilizes the difference frequency effect of nonlinear crystals. When a femtosecond laser pulse is incident, it performs difference frequency operations among the various frequency components of the laser, thereby generating terahertz-band radiation. However, both of these methods share a common problem: they are limited by the semiconductor breakdown threshold and crystal damage threshold, and cannot simply increase the terahertz radiation energy by enhancing the laser power.

On the other hand, electron beams can produce electromagnetic radiation in various ways: transition radiation occurs when an electron beam passes through an interface composed of two media with different dielectric constants; synchrotron radiation occurs when the direction of electron velocity changes; Cherenkov radiation occurs when electrons propagate in a transparent medium, etc. SLAC obtained terahertz radiation with a single-pulse energy exceeding 600μJ through coherent transition radiation in 2013 [60]. This type of method generally relies on traditional accelerators, and thus all have the disadvantages of large size and high cost, which are not conducive to promotion and development.

Using high-energy electrons obtained from laser-solid target interactions to produce terahertz radiation can effectively convert high-power lasers into terahertz radiation while maintaining a small size. Laser-solid target interactions generate hot electrons, which form a sheath field and cause ion acceleration when they propagate forward and pass through the target back surface. At the same time, electrons produce radiation through various mechanisms: some electrons directly penetrate the surface and continue to propagate, forming transition radiation; some electrons penetrate the surface and then decelerate and reverse direction under the influence of the target back sheath field, forming bremsstrahlung radiation; meanwhile, the target back sheath field, as a charge separation field, decays over time, and radiation also occurs in this process. For femtosecond laser-driven electron beams, the main radiation generation mechanism is usually transition radiation. When electrons pass through an interface composed of two media with different dielectric constants, the discontinuity of the electric field at the interface generates a time-varying polarization current, which then radiates electromagnetic waves outward. This is transition radiation. The energy-angle distribution of the radiation is related to the electron energy and incidence angle, with higher incident electron energy resulting in better collimation of the radiation. For femtosecond laser-driven electron beams, the electron beam pulse width is approximately equal to the laser pulse width, which is less than the terahertz wavelength. Therefore, the radiation is usually coherent transition radiation, with the radiation intensity being proportional to N^2, where N is the number of electrons. Experimentally, by interacting a picosecond laser with an energy of 60 J with a metal solid target, terahertz radiation with a total energy as high as 55 mJ can be obtained, with an energy conversion efficiency of about 0.1% [61].

For traditional solid target ion acceleration, since the plasma formed on the front surface of the solid target has a steep density gradient, most of the laser energy is reflected by the solid target, and only a small portion of the energy is transferred to the hot electrons, resulting in relatively low energy conversion efficiency from laser to terahertz radiation. To solve this problem, a feasible solution is to use an ablation laser to create a large-scale low-density pre-plasma on the front surface of the target, thereby increasing the energy deposition from the laser to the electrons and thus improving the overall energy conversion efficiency [62]. The schematic diagram of this scheme is shown in Figure 10.13. The specific process is as follows: first, a low-energy ablation laser is used to strike the front surface of the solid target, causing the target to expand forward and form a large-scale low-density pre-plasma; then the main laser pulse interacts with the pre-plasma to generate a large number of ultra-hot electrons; when the electrons pass through the target back surface, terahertz waves are generated through transition radiation.

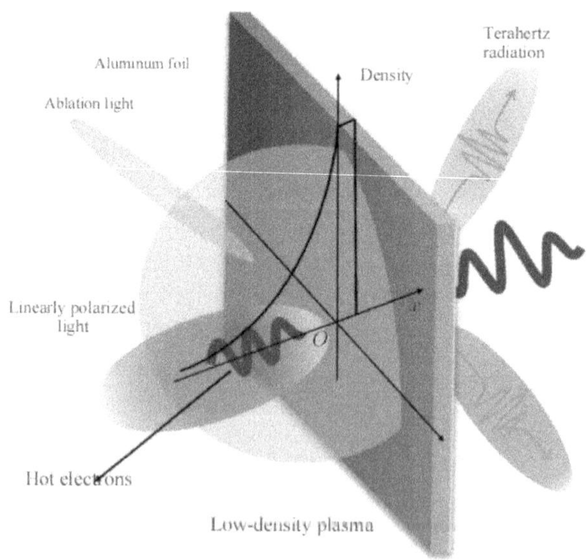

Fig. 10.13 Schematic diagram of the pre-ablation gain terahertz radiation scheme [62].

In the laser-aluminum target interaction experiment, as shown in Figure 10.14, compared with the case without ablation light, a significant gain in terahertz radiation was observed in the aluminum target with ablation light acting 100 ps in advance, which is consistent with the theoretical and simulation predictions. By adjusting the delay between the ablation light and the main laser, the experiment found that there exists an optimal ablation time for terahertz radiation generation. For an aluminum target with a thickness of μm, the optimal ablation time is around 100 ps. In addition, the experiment also investigated the electromagnetic pulse (EMP) radiation generated by high-energy electrons produced by target front ablation on the target back surface, and observed a significant gain in EMP due to ablation.

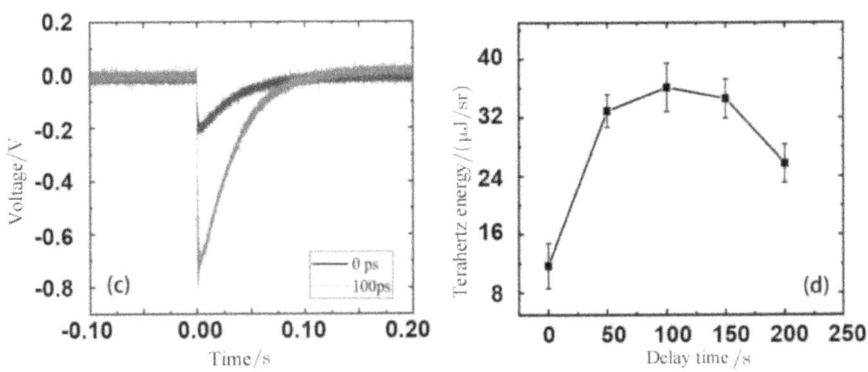

Fig. 10.14 Terahertz radiation yield of aluminum targets under different ablation times [62].

10.2.2 *Extreme Ultraviolet Radiation*

Extreme ultraviolet radiation is a concept in spectroscopy, generally referring to the wavelength band between soft X-rays and deep ultraviolet radiation. There is no very strict definition of the wavelength range of extreme ultraviolet radiation. Traditionally, radiation with wavelengths between 10 and 124 nm is considered extreme ultraviolet radiation. Since the wavelength of extreme ultraviolet radiation is relatively short and the corresponding photon energy exceeds 10 eV, it is not commonly seen in daily life. The extreme ultraviolet radiation existing in nature mainly comes from the solar corona. The temperature of the corona is as high as 10^6 K, which is sufficient to produce thermal radiation in the extreme ultraviolet band.

With the advancement of technological means, humans have now been able to obtain extreme ultraviolet radiation far brighter than that of the sun. The extreme ultraviolet radiation generated in laboratories is mainly divided into two types. One is extreme ultraviolet radiation based on accelerator electron beams, typically synchrotron radiation and free electron lasers (FEL). The other is extreme ultraviolet radiation based on plasma, including coherent sources such as extreme ultraviolet lasers and high harmonic generation (HHG), and incoherent sources such as plasma emission. High harmonics generally originate from the interaction of lasers with media, and can be divided into gas high harmonics and plasma surface high harmonics according to the intensity of the driving laser. Plasma emission can be divided into laser-produced plasma (LPP) and discharge-produced plasma (DPP) according to the driving source.

Utilizing ultra-intense femtosecond pulse lasers to accelerate heavy ions has always been a challenge in the field of laser acceleration. In previous experimental studies, heavy ions are usually only accelerated to a maximum energy of several MeV per nucleon (MeV/u). In contrast, protons can be accelerated to nearly 100 MeV under the same conditions, much higher than that of heavy ions. This is because to effectively accelerate heavy ions, it is necessary to ionize them to a high charge state at the initial stage of acceleration and inject them into the acceleration field, while maintaining a sufficiently long acceleration time. Generally, it is difficult to achieve both conditions simultaneously. Experimental research has found that by using a double-layer composite target composed of ultra-thin, ultra-low-density carbon nanotube foam and diamond-like carbon (DLC) nanofilm, both of these conditions can be successfully achieved. Figure 10.15 shows the electron microscope image of the carbon nanotube foam. This carbon nanotube foam uses carbon nanotube fibers with diameters of only tens of nanometers to achieve a relatively uniform critical density plasma at the laser focal spot scale. When this composite target interacts with an ultra-intense femtosecond pulse laser, the carbon ions in the DLC film successively undergo photoionization injection and sheath field acceleration over a period of hundreds of femtoseconds, ultimately reaching a speed of 30% of the speed of light.

During ion acceleration, high-efficiency secondary extreme ultraviolet radiation can also be measured experimentally. As shown in Figure 10.16, the interaction of relativistic lasers with carbon nanotube foam targets can produce broadband extreme ultraviolet radiation. Comparing with the atomic radiation spectrum lines, it can be found that there are many

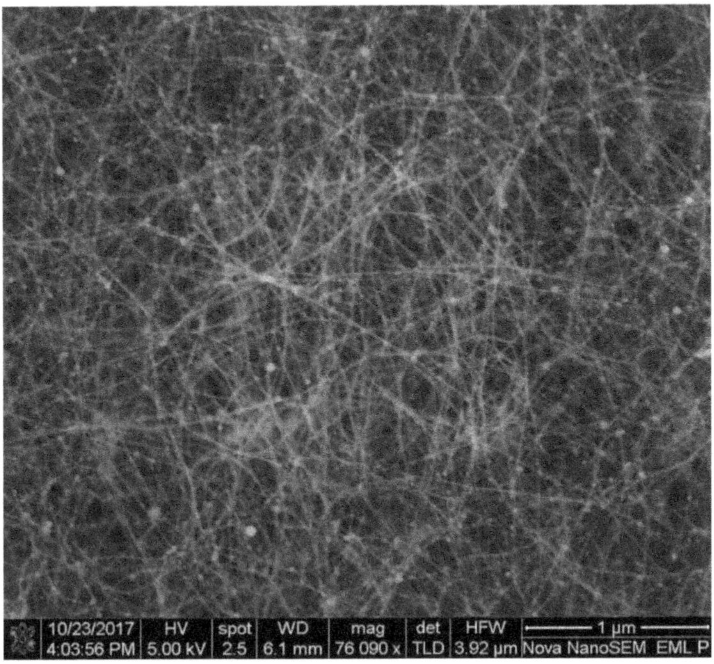

Fig. 10.15 Electron microscope image of carbon nanotube foam.

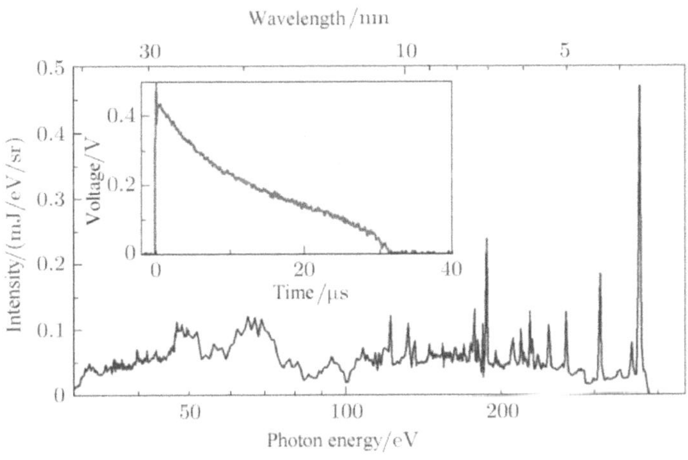

Fig. 10.16 Extreme ultraviolet radiation spectrum produced by a relativistic laser interacting with a 60 μm carbon nanotube foam target. The inset shows the radiation energy signal measured by an AXUV diode at a 15° angle in the same experimental shot.

line radiations in the short-wave band of this extreme ultraviolet radiation, which mainly come from carbon and iron elements in the carbon nanotube foam target. In the 3 to 7 nm wavelength band, the line radiation mainly comes from C^{5+} and Fe^{15+} ions. According to the NIST database, the ionization energies required for both are 490.0 eV and 489.3 eV, respectively, which are very close, so it is reasonable that iron atoms are ionized to the 15th charge state. The charge states of carbon ions and iron ions here are also consistent with the field ionization charge states at similar laser intensities reported in previous literature [63].

Overall, through measurements of the spectrum, energy, and source size, the experimental findings regarding the secondary extreme ultraviolet radiation from ion acceleration based on carbon nanotube foam targets are as follows: First, the interaction of relativistic lasers with carbon nanotube foam targets can produce broadband extreme ultraviolet radiation. The spectral distribution of this extreme ultraviolet radiation is relatively flat, with a higher proportion of line radiation in the wavelength band below 10 nm, mainly from carbon and iron elements, and mainly continuous radiation in the wavelength band above 10 nm. Second, this extreme ultraviolet radiation has a certain degree of forwardness. Considering the solid angle, the total radiation yield is 240 mJ, corresponding to a conversion efficiency of 20%. The upper limit of the source size for this extreme ultraviolet radiation is 120 μm. Assuming that the main energy of this extreme ultraviolet radiation is distributed within a 10 ps time window, taking radiation with a photon energy of 70 eV as an example, the brightness can be calculated to be on the order of 10^{20} photons/s/mm^2/mrad^2/0.1%BW, comparable to that of traditional third-generation synchrotron radiation sources.

This secondary extreme ultraviolet radiation has many application prospects. The energy of extreme ultraviolet radiation is moderate, corresponding to the transition energy levels of outer or middle shell electrons of atoms. Therefore, it can be used to detect the electronic energy level structure and chemical bond composition at the atomic level by coupling extreme ultraviolet radiation with outer or middle shell electrons, thereby providing a completely new means for studying microscopic substances and helping to understand the physical reasons behind macroscopic material properties. For the same reason, extreme ultraviolet radiation can also be used for microscopy to achieve a resolution of tens of nanometers, better resolving the structure of materials. Based on the same principle, extreme ultraviolet radiation may also be applicable to nanofabrication and high-precision lithography.

10.2.3 *X-ray Radiation*

Compared with synchrotron radiation sources, femtosecond laser-driven X-ray sources have a series of advantages such as high brightness, short pulse width, low facility cost, and small facility footprint, and have broad application prospects in fields such as material science, transmission imaging, and national defense security inspection. The currently mature femtosecond laser-driven ultra-short X-ray sources all use low-density gas targets, and the conversion efficiency from laser to X-ray is in the range of $10^{-7} \sim 10^{-5}$, which limits their further applications. Compared with gas targets, a series of theoretical studies have shown that using steep-edged, uniform critical density plasma can simultaneously achieve a large number of directly laser-accelerated (DLA) electrons and a strong self-generated lateral driving magnetic field, thereby realizing high-efficiency X-ray radiation. The physical picture is shown in Figure 10.17: The laser propagates forward in the critical density plasma to form a channel, while simultaneously undergoing nonlinear effects such as relativistic self-focusing and self-steepening. At the same time, electrons in the plasma are subjected to the ponderomotive force of the laser field and are accelerated forward. Since the electron density in the critical density plasma is high, these accelerated electrons form a very strong longitudinal current and a toroidal self-generated magnetic field. The toroidal magnetic field further confines the electrons in the channel region, enabling the electrons to undergo direct acceleration processes including resonant acceleration, forming high-energy electrons with energy exceeding the ponderomotive potential. This process is also accompanied by the cyclotron motion of electrons under the combined action of

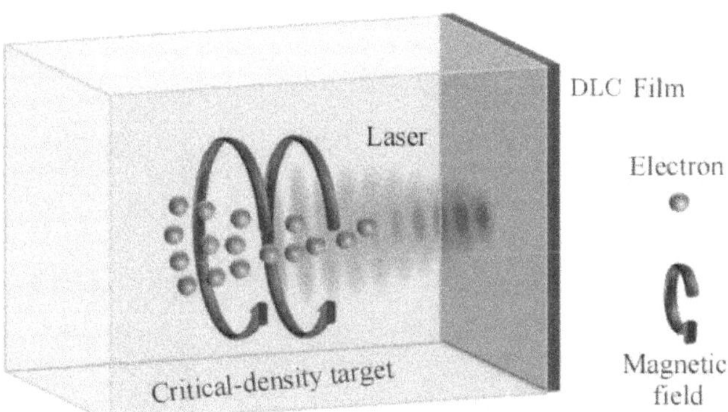

Fig. 10.17 X-ray radiation in critical density plasma.

the charge separation field, self-generated magnetic field, and laser field. This type of electron motion similar to synchrotron radiation can very effectively produce X-rays. In addition, if a second layer of diamond-like carbon (DLC) film target is used to reflect the laser and directly accelerated electrons for collision, higher-energy X-rays and γ-rays can be obtained through the inverse Compton scattering mechanism.

Obtaining uniform critical density plasma in the laboratory is not easy. Taking a common 800 nm laser as an example, the corresponding critical density n_c is $1.8 \times 10^{21}/\text{cm}^3$. The density of common solid materials is generally in the g/cm^3 range, and the corresponding plasma density in the fully ionized state is $N_e = \rho Z N_A A$. Here, $N_A = 6.02 \times 10^{23}/\text{mol}$ is Avogadro's constant, and Z and A are the average ionization degree and average atomic mass number, respectively. Therefore, for common solids, the corresponding plasma density is hundreds of n_c, and they cannot be used directly as critical density target materials. As for gas nozzles, which are commonly used as electron acceleration target materials, the plasma density they can generate is generally around $0.01n_c$, and they are usually not suitable as critical density target materials.

To obtain critical density plasma in experiments, various attempts have been made. Experimental research has found that using carbon nanotubes as critical density target materials can achieve high-efficiency ion acceleration while obtaining high-efficiency X-ray radiation. Using a PW laser to interact with carbon nanotube targets, the conversion efficiency from laser to X-ray can be experimentally increased to 1%, with an average X-ray energy reaching hundreds of keV. This unprecedented high-efficiency femtosecond X-ray source will greatly inspire related research work.

In addition, theoretical and simulation studies have found that the photon energy obtained using carbon nanotube targets is proportional to the cube of the light intensity. Based on this scaling law, if a 10 PW laser currently under construction is used to interact with carbon nanotube targets, it is expected to obtain a high-efficiency γ-ray source, thereby promoting research in photonicuclear reactions and laboratory astrophysics.

References

[1] Patel P. K., Mackinnon A. J., Key M. H., *et al.* Isochoric heating of solid-density matter with an ultrafast proton beam [J]. *Physical Review Letters*, 2003, 91(12): 125004.

[2] Gauthier M., Chen S. N., Levy A., *et al.* Charge equilibrium of a laser-generated carbon-ion beam in warm dense matter [J]. *Physical Review Letters*, 2013, 110(13): 135003.
[3] Vozenin M. C., Hendry J. H., Limoli C. L. Biological benefits of ultra-high dose rate FLASH radiotherapy: sleeping beauty awoken [J]. *Clinical Oncology*, 2019, 31(7): 407–415.
[4] Wilson J. D., Hammond E. M., Higgins G. S., *et al.* Ultra-high dose rate (FLASH) radiotherapy: silver bullet or fool's gold? [J]. *Journal of Radiation Research*, 2020, 9(2020): 1563.
[5] Chaffer Christine L., Weinberg Robert A. A perspective on cancer cell metastasis [J]. *Science*, 2011, 331(6024): 1559–1564.
[6] Steeg P. S. Targeting metastasis [J]. *Nature Reviews Cancer*, 2016, 16(4): 201–218.
[7] Boustani J., Grapin M., Laurent P.-A., *et al.* The 6th R of radiobiology: reactivation of anti-tumor immune response [J]. *Cancers*, 2019, 11(6): 860.
[8] Mole R. H. Whole body irradiation-radiobiology or medicine? [J]. *The British Journal of Radiology*, 1953, 26(305): 234–241.
[9] Mclaughlin M., Patin E. C., Pedersen M., *et al.* Inflammatory microenvironment remodelling by tumour cells after radiotherapy [J]. *Nature Reviews Cancer*, 2020, 20(4): 203–217.
[10] Yang G., Wu L., Chen L., *et al.* Targeted irradiation of shoot apical meristem of arabidopsis embryos induces long-distance bystander/abscopal effects [J]. *Radiation Research*, 2007, 167(3): 298–305.
[11] Yang G., Mei T., Yuan H., *et al.* Bystander/abscopal effects induced in intact arabidopsis seeds by low-energy heavy-ion radiation [J]. *Radiation Research*, 2008, 170(3): 372–380.
[12] Yang G., Wu L., Chen S., *et al.* Mitochondrial dysfunction resulting from loss of cytochrome c impairs radiation-induced bystander effect [J]. *British Journal of Cancer*, 2009, 100(12): 1912–1916.
[13] Yang G., Quan Y., Wang W., *et al.* Dynamic equilibrium between cancer stem cells and non-stem cancer cells in human SW620 and MCF-7 cancer cell populations [J]. *British Journal of Cancer*, 2012, 106(9): 1512–1519.
[14] Liu Y., Kobayashi A., Maeda T., *et al.* Target irradiation induced bystander effects between stem-like and non stem-like cancer cells [J]. *Mutation Research/Fundamental and Molecular Mechanisms of Mutagenesis*, 2015, 773(2015): 43–47.
[15] Liu Y., Kobayashi A., Fu Q., *et al.* Rescue of targeted nonstem-like cells from bystander stem-like cells in human fibrosarcoma HT1080 [J]. *Radiation Research*, 2015, 184(3): 334–340.
[16] Quan Q., Wang X., Lu C., *et al.* Association of extracellular matrix microarchitecture and three-dimensional collective invasion of cancer cells [J]. *Biotechnic & Histochemistry*, 2020, 95(8): 605–612.
[17] Lehrack S., Assmann W., Parodi K. Ionoacoustics for range monitoring of proton therapy [J]. *Journal of Physics: Conference Series*, 2019, 1154(2019): 012015.
[18] Bisogni M. G., Attili A., Battistoni G., *et al.* Inside in-beam positron

emission tomography system for particle range monitoring in hadrontherapy [J]. *Journal of Medical Imaging*, 2017, 4(1): 011005.
[19] Polf J. C., Parodi K. Imaging particle beams for cancer treatment [J]. *Physics Today*, 2015, 68(10): 28–33.
[20] Jones K. C., Nie W., Chu J. C. H., *et al.* Acoustic-based proton range verification in heterogeneous tissue: simulation studies [J]. *Phys. Med. Biol.*, 2018, 63(2): 025018.
[21] Parodi K., Assmann W. Ionoacoustics: a new direct method for range verification [J]. *Modern Physics Letters A*, 2015, 30(17): 1540025.
[22] Sulak L., Armstrong T., Baranger H., *et al.* Experimental studies of the acoustic signature of proton-beams traversing fluid media [J]. *Nucl. Instrum. Methods*, 1979, 161(2): 203–217.
[23] Hayakawa J. T. Y., Arai N., Hosono K., *et al.* Acoustic pulse generated in a patient during treatment by pulsed proton radiation beam [J]. *Radiation Oncology Investigations*, 1995, 3(1): 42–45.
[24] Assmann W., Kellnberger S., Reinhardt S., *et al.* Ionoacoustic characterization of the proton Bragg peak with submillimeter accuracy [J]. *Med. Phys.*, 2015, 42(2): 567–574.
[25] Kellnberger S., Assmann W., Lehrack S., *et al.* Ionoacoustic tomography of the proton Bragg peak in combination with ultrasound and optoacoustic imaging [J]. *Sci. Rep.*, 2016, 6(1): 29305.
[26] Lehrack S., Assmann W., Bertrand D., *et al.* Submillimeter ionoacoustic range determination for protons in water at a clinical synchrocyclotron [J]. *Phys. Med. Biol.*, 2017, 62(17): L20–L30.
[27] Haffa D., Yang R., Bin J., *et al.* I-BEAT: ultrasonic method for online measurement of the energy distribution of a single ion bunch [J]. *Sci. Rep.*, 2019, 9(1): 6714.
[28] Jones K. C., Witztum A., Sehgal C. M., *et al.* Proton beam characterization by proton-induced acoustic emission: simulation studies [J]. *Phys. Med. Biol.*, 2014, 59(21): 6549–6563.
[29] Jones K. C., Vander Stappen F., Sehgal C. M., *et al.* Acoustic time-of-flight for proton range verification in water [J]. *Med. Phys.*, 2016, 43(9): 5213.
[30] Patch S. K., Kireeff Covo M., Jackson A., *et al.* Thermoacoustic range verification using a clinical ultrasound array provides perfectly co-registered overlay of the Bragg peak onto an ultrasound image [J]. *Phys. Med. Biol.*, 2016, 61(15): 5621–5638.
[31] Patch S. K., Hoff D. E. M., Webb T. B., *et al.* Two-stage ionoacoustic range verification leveraging Monte Carlo and acoustic simulations to stably account for tissue inhomogeneity and accelerator-specific time structure - a simulation study [J]. *Med. Phys.*, 2018, 45(2): 783–793.
[32] Rohrer L., Jakob H., Rudolph K., *et al.* The four gap double drift buncher at Munich [J]. *Nuclear Instruments and Methods in Physics Research*, 1984, 220(1): 161–164.
[33] Jones K. C., Vander Stappen F., Bawiec C. R., *et al.* Experimental observation of acoustic emissions generated by a pulsed proton beam from a hospital-based clinical cyclotron [J]. *Med. Phys.*, 2015, 42(12): 7090–7097.

[34] Borghesi M., Campbell D. H., Schiavi A., *et al.* Electric field detection in laser-plasma interaction experiments via the proton imaging technique [J]. *Physics of Plasmas*, 2002, 9(5): 2214–2220.
[35] Kar S., Ahmed H., Prasad R., *et al.* Guided post-acceleration of laser-driven ions by a miniature modular structure [J]. *Nature Communications*, 2016, 7(1): 10792.
[36] Barberio M., Vallieres S., Sciscio M., *et al.* Graphitization of diamond by laser-accelerated proton beams [J]. *Carbon*, 2018, 139(2018): 531–537.
[37] Barberio M., Sciscio M., Vallieres S., *et al.* Laser-accelerated particle beams for stress testing of materials [J]. *Nature Communications*, 2018, 9(1): 372.
[38] Barberio M., Sciscio M., Vallieres S., *et al.* Laser-generated proton beams for high-precision ultra-fast crystal synthesis [J]. *Scientific Reports*, 2017, 7(1): 12522.
[39] Barberio M., Giusepponi S., Vallières S., *et al.* Ultra-fast high-precision metallic nanoparticle synthesis using laser-accelerated Protons [J]. *Scientific Reports*, 2020, 10(1): 9570.
[40] Tabak M., Hammer J., Glinsky M. E., *et al.* Ignition and high gain with ultrapowerful lasers [J]. *Physics of Plasmas*, 1994, 1(5): 1626–1634.
[41] Kodama R., Shiraga H., Shigemori K., *et al.* Fast heating scalable to laser fusion ignition [J]. *Nature*, 2002, 418(6901): 933–934.
[42] Azechi H., Mima K., Shiraga S., *et al.* Present status of fast ignition realization experiment and inertial fusion energy development [J]. *Nuclear Fusion*, 2013, 53(10): 104021.
[43] Theobald W., Solodov A., Stoeckl C., *et al.* Initial cone-in-shell fast-ignition experiments on OMEGA [J]. *Physics of Plasmas*, 2011, 18(5): 056305.
[44] Hand E. Laser fusion nears crucial milestone [J]. *Nature News*, 2012, 483(7388): 133.
[45] Putvinski S., Ryutov D., Yushmanov P. Fusion reactivity of the pB11 plasma revisited [J]. *Nuclear Fusion*, 2019, 59(7): 076018.
[46] Labaune C., Baccou C., Depierreux S., *et al.* Fusion reactions initiated by laser-accelerated particle beams in a laser-produced plasma [J]. *Nature Communications*, 2013, 4(1): 2506.
[47] Picciotto A., Margarone D., Velyhan A., *et al.* Boron-broton nuclear-fusion enhancement induced in boron-doped silicon targets by low-contrast pulsed laser [J]. *Physical Review X*, 2014, 4(3).
[48] Giuffrida L., Belloni F., Margarone D., *et al.* High-current stream of energetic alpha particles from laser-driven proton-boron fusion [J]. *Phys. Rev. E.*, 2020, 101(1-1): 013204.
[49] Liao G., Li Y., Liu H., *et al.* Multimillijoule coherent terahertz bursts from picosecond laser-irradiated metal foils [J]. *Proceedings of the National Academy of Sciences*, 2019, 116(10): 3994–3999.
[50] Nie Z., Pai C.-H., Hua J., *et al.* Relativistic single-cycle tunable infrared pulses generated from a tailored plasma density structure [J]. *Nature Photonics*, 2018, 12(8): 489–494.

[51] Yeung M., Rykovanov S., Bierbach J., *et al.* Experimental observation of attosecond control over relativistic electron bunches with two-colour fields [J]. *Nature Photonics*, 2017, 11(1): 32–35.
[52] Huang K., Li M. H., Yan W. C., *et al.* Intense high repetition rate Mo K? X-ray source generated from laser solid interaction for imaging application [J]. *Review of Scientific Instruments*, 2014, 85(11): 113304.
[53] Cipiccia S., Islam M. R., Ersfeld B., *et al.* Gamma-rays from harmonically resonant betatron oscillations in a plasma wake [J]. *Nature Physics*, 2011, 7(11): 867–871.
[54] Ta Phuoc K., Corde S., Thaury C., *et al.* All-optical Compton gamma-ray source [J]. *Nature Photonics*, 2012, 6(5): 308–311.
[55] Subiel A. Feasibility studies on the application of relativistic electron beams from a laser plasma wakefield accelerator in radiotherapy [D]. Glasgow: University of Strathclyde, 2014.
[56] West D., Sherwood A. C. Radiography with 160 MeV protons [J]. *Nature*, 1972, 239(5368): 157–159.
[57] Wang W., Shen B., Zhang H., *et al.* Large-scale proton radiography with micrometer spatial resolution using femtosecond petawatt laser system [J]. *AIP Advances*, 2015, 5: 107214.
[58] Roth M., Cowan T. E., Key M. H., *et al.* Fast ignition by intense laser-accelerated proton beams [J]. *Physical Review Letters*, 2001, 86(3): 436.
[59] Taichiro G. Radiation as an in situ auto-vaccination: current perspectives and challenges [J]. *Vaccines*, 2019, 7(3): 100.
[60] Wu Z., Fisher A. S., Goodfellow J., *et al.* Intense terahertz pulses from SLAC electron beams using coherent transition radiation [J]. *Review of Scientific Instruments*, 2013, 84(2): 022701.
[61] Liao G., Li Y., Liu H., Scott G. G., *et al.* Multimillijoule coherent terahertz bursts from picosecond laser-irradiated metal foils [J]. *Proceedings of the National Academy of Sciences of the United States of America*, 2019, 116: 3994–3999.
[62] Geng Y., Li D., Zhang S., *et al.* Strong enhancement of coherent terahertz radiation by target ablation using picosecond laser pulses [J]. *Physics of Plasmas*, 2020, 27(11): 113104.
[63] Shou Y. R., Pan Z., Cao Z. X., *et al.* Efficient broadband extreme ultraviolet radiation based on carbon nanotube foam [J]. *Chinese Journal of Optics*, 2022, 42(11): 257–263.